VOLCANOES

Other books in the Great Disasters series:

The Black Death
Earthquakes
Tornadoes

GREAT DISASTERS

VOLCANOES

Nancy Harris and Lynn Armstrong, *Book Editors*

Daniel Leone, *President*
Bonnie Szumski, *Publisher*
Scott Barbour, *Managing Editor*

**GREENHAVEN
PRESS** ®

San Diego • Detroit • New York • San Francisco • Cleveland
New Haven, Conn. • Waterville, Maine • London • Munich

For more information, contact
Greenhaven Press
27500 Drake Rd.
Farmington Hills, MI 48331-3535
Or you can visit our Internet site at http://www.gale.com

LIBRARY OF CONGRESS CATALOGING-IN-PUBLICATION DATA

Volcanoes / Nancy Harris, book editor, Lynn Armstrong, book editor.
    p. cm. — (Great disasters)
Includes bibliographical references and index.
ISBN 0-7377-1439-5 (lib. bdg. : alk. paper) —
ISBN 0-7377-1440-9 (pbk. : alk. paper)
    1. Volcanoes. I. Harris, Nancy. II. Armstrong, Lynn. III. Great disasters
(Greenhaven Press)
QE521 .V67 2003
363.34'95—dc21   2002072368

CONTENTS

Chapter 2: Volcanic Disasters

H umans have an ambivalent relationship with their home planet, nurtured on the one hand by Earth's bounty but devastated on the other hand by its catastrophic natural disasters. While these events are the results of the natural processes of Earth, their consequences for humans frequently include the disastrous destruction of lives and property. For example, when the volcanic island of Krakatau exploded in 1883, the eruption generated vast seismic sea waves called tsunamis that killed about thirty-six thousand people in Indonesia. In a single twenty-four-hour period in the United States in 1974, at least 148 tornadoes carved paths of death and destruction across thirteen states. In 1976, an earthquake completely destroyed the industrial city of Tangshan, China, killing more than 250,000 residents.

Some natural disasters have gone beyond relatively localized destruction to completely alter the course of human history. Archaeological evidence suggests that one of the greatest natural disasters in world history happened in A.D. 535, when an Indonesian "supervolcano" exploded near the same site where Krakatau arose later. The dust and debris from this gigantic eruption blocked the light and heat of the sun for eighteen months, radically altering weather patterns around the world and causing crop failure in Asia and the Middle East. Rodent populations increased with the weather changes, causing an epidemic of bubonic plague that decimated entire populations in Africa and Europe. The most powerful volcanic eruption in recorded human history also happened in Indonesia. When the volcano Tambora erupted in 1815, it ejected an estimated 1.7 million tons of debris in an explosion that was heard more than a thousand miles away and that continued to rumble for three months. Atmospheric dust from the eruption blocked much of the sun's heat, producing what was called "the year without summer" and creating worldwide climatic havoc, starvation, and disease.

As these examples illustrate, natural disasters can have as much impact on human societies as the bloodiest wars and most chaotic political revolutions. Therefore, they are as worthy of study as the major events of world history. As with the study of social and political events, the exploration of natural disasters can illuminate the causes of these catastrophes and target the lessons learned about how to mitigate and prevent the loss of life when disaster strikes again. By examining these events and the forces behind them, the Greenhaven Press Great Disasters series is designed to help students better understand such cataclysmic events. Each anthology in the series focuses on a specific type of natural disaster or a particular disastrous event in history. An introductory essay provides a general overview of the subject of the anthology, placing natural disasters in historical and scientific context. The essays that follow, written by specialists in the field, researchers, journalists, witnesses, and scientists, explore the science and nature of natural disasters, describing particular disasters in detail and discussing related issues, such as predicting, averting, or managing disasters. To aid the reader in choosing appropriate material, each essay is preceded by a concise summary of its content and biographical information about its author.

In addition, each volume contains extensive material to help the student researcher. An annotated table of contents and a comprehensive index help readers quickly locate particular subjects of interest. To guide students in further research, each volume features an extensive bibliography including books, periodicals, and related Internet websites. Finally, appendixes provide glossaries of terms, tables of measurements, chronological charts of major disasters, and related materials. With its many useful features, the Greenhaven Press Great Disasters series offers students a fascinating and awe-inspiring look at the deadly power of Earth's natural forces and their catastrophic impact on humans.

INTRODUCTION

I n April 1815 the lieutenant governor of Java sent two boats out to investigate what sounded like cannon fire. Tambora, a volcano on the East Indian island of Sumbawa, Indonesia, was blowing its top. It ejected 1.7 million tons of debris, hurling rocks at the town of Saugar twenty-five miles away. Tambora's eruption was heard more than a thousand miles away, and it continued to rumble for three months. The volcanic mountain was flattened into a broad plateau beneath a two-foot layer of ash and mud. The surrounding sea was jammed with thousands of uprooted trees and huge blocks of pumice (sponge-textured volcanic rock), which ships would be forced to navigate around for the next year.

The eruption of Tambora created worldwide climatic havoc, starvation, and disease. Most of the volcanic debris fell back onto the island of Sumbawa, but the remainder was pulverized into dust and carried into the upper atmosphere to travel around the earth. This atmospheric dust blocked some of the sun's heat, creating what was called "the year without summer." From Maine to Connecticut, it snowed in June. On Sumbawa and the nearby islands, 10,000 people were killed by the eruption and another 82,000 died later due to disease and famine; this total of 92,000 deaths was the largest ever recorded for a volcanic disaster.

Another Indonesian volcano, Mount Pelée, on the island of Martinique, exploded in May 1902. The eruption blasted away the side of the mountain, sending a deadly cloud of red-hot gas down the mountainside, burning everything in its path. In three minutes, the inferno killed twenty-eight thousand people, almost the entire population of the city of St. Pierre.

In November 1985 an ice cap exploded above the town of Armero, Colombia. Scientists had issued a warning when the high volcano had begun rumbling but the residents of the town either ignored it or did not get the warning in time. The volcanic eruption melted snowfields that picked up debris, and a monstrous mudflow ensued that engulfed villages thirty and forty miles away, eventually killing twenty-six thousand people.

As these examples illustrate, volcanoes represent a significant threat to human life and property. Volcanoes have been erupting on the planet since the beginning of the earth's formation, and they can be as unpredictable as the weather. After centuries, and even millennia, volcanic mountains thought to be extinct can awaken and erupt either uneventfully or with great violence. Some erupt daily. In the history of volcanic eruptions, two-thirds have caused fatalities. During the past four hundred years, five hundred volcanoes have erupted, killing more than two hundred thousand people. During the twentieth century, an average of

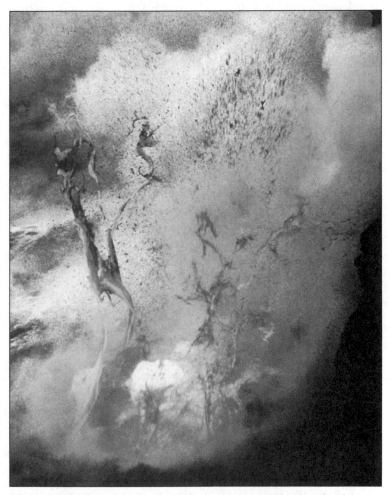

The unpredictable and violent force of volcanic eruptions has posed a threat to human life and property for thousands of years.

more than eight hundred people per year died due to volcanic disaster. A single volcanic eruption can cause property damage amounting to hundreds of millions of dollars, and a very powerful eruption occurring in or near a heavily populated area might cause billions of dollars of damage.

The eruption of Mount St. Helens in 1980, the largest eruption in the continental United States for several centuries, killed fifty-seven people and began a decade of increased volcanic activity worldwide. Since then, more deaths have occurred. Because of the population growth pushing people into closer proximity to volcanoes and the public and governmental lack of awareness or disregard of the potential danger, more lives are increasingly at risk. As a result, averting disaster has become a high priority for volcanologists.

Hoping to prevent such extreme destruction, scientists are attempting to better understand how volcanoes work so they can improve their ability to predict eruptions. During the past century, researchers have made significant progress in understanding the nature of volcanoes and in developing new monitoring techniques. They have mapped sites where magma has risen to the earth's surface and studied the geologic record of previous eruptions. Research has shown that eruptions are preceded by movements of magma from deep internal reservoirs upward toward the earth's surface. Warning signals that usually accompany this movement include earthquake tremors, land deformations (changes in the elevation of land surrounding the volcano), and gas emissions. Modern volcanologists have developed technology to monitor these signals, yet much work remains to be done. Currently, for instance, only about 150 out of 1,000 volcanic danger zones are being monitored.

Mount Pinatubo may be considered a scientific success story. In June 1991, Mount Pinatubo in the Philippines was responsible for what may be the largest eruption of the century. It launched 20 million tons of sulfur dioxide twenty-two miles into the atmosphere, altering weather and temperatures worldwide. During the first three months, the eruption killed at least seven hundred people and left tens of thousands homeless. However, scientists averted a more catastrophic disaster by predicting the eruption and warning the public in time to evacuate fifteen thousand people from the immediate area.

Studies in volcanology have also revealed that, despite their short-term destructive power, volcanoes have provided many benefits to the ecological systems of the earth. Through geologic time, volcanoes have created much of the planet's land surface and atmosphere. The breakdown of volcanic rock and debris has created some of the world's most fertile soils, which is partly why so many people have settled near volcanoes. Recently, in places of volcanic cataclysm, scientists have been amazed to witness the rapid rejuvenation of plant and animal life there. Volcanoes are a natural part of the earth's life cycle. It behooves humankind to learn to live safely with them.

The Science and Study of Volcanoes

Physical Features of Volcanoes

By Robert I. Tilling

Robert I. Tilling has worked as a volcanologist for thirty years. Tilling's many positions with the U.S. Geological Survey (USGS) have included scientist-in-charge at the Hawaiian Volcano Observatory (HVO) and chief scientist of the Volcano Hazards Team.

This excerpt from Tilling's USGS publication Volcanoes *describes the physical characteristics of volcanoes, the different types of volcanoes, and the locations of the most typical volcanic environments. In his explanation of extraterrestrial volcanism, the author relates how the* Voyager 2 *spacecraft captured images of active volcanism on Io, one of Jupiter's moons. He also describes submarine volcanoes and volcanic vents, which spew steam and rock debris high above the ocean's surface. He details the process through which the combination of hot lava and seawater creates the well-known black sand beaches of Hawaii. Tilling also details the benefits that volcanoes have for humans and warns that, despite advances that have been made in volcanology, improvements are needed in predicting eruptions and warning the public.*

V olcanoes destroy and volcanoes create. The catastrophic eruption of Mount St. Helens on May 18, 1980, made clear the awesome destructive power of a volcano. Yet, over a time span longer than human memory and record, volcanoes have played a key role in forming and modifying the planet upon which we live. More than 80 percent of the Earth's surface—above and below sea level—is of volcanic origin. Gaseous emissions from volcanic vents over hundreds of millions of years formed the Earth's earliest oceans and atmosphere, which supplied the ingredients vital to evolve and sustain life. Over geologic eons, countless volcanic eruptions have produced mountains,

Robert I. Tilling, *Volcanoes*. Washington, DC: United States Geological Survey, 1999.

plateaus, and plains, which subsequent erosion and weathering have sculpted into majestic landscapes and formed fertile soils.

Ironically, these volcanic soils and inviting terrains have attracted, and continue to attract, people to live on the flanks of volcanoes. Thus, as population density increases in regions of active or potentially active volcanoes, mankind must become increasingly aware of the hazards and learn not to "crowd" the volcanoes. People living in the shadow of volcanoes must live in harmony with them and expect, and should plan for, periodic violent unleashings of their pent-up energy. . . .

On August 24, A.D. 79, Vesuvius Volcano suddenly exploded and destroyed the Roman cities of Pompeii and Herculaneum. Although Vesuvius had shown stirrings of life when a succession of earthquakes in A.D. 63 caused some damage, it had been literally quiet for hundreds of years and was considered "extinct." Its surface and crater were green and covered with vegetation, so the eruption was totally unexpected. Yet in a few hours, hot volcanic ash and dust buried the two cities so thoroughly that their ruins were not uncovered for nearly 1,700 years, when the discovery of an outer wall in 1748 started a period of modern archeology. Vesuvius has continued its activity intermittently ever since A.D. 79 with numerous minor eruptions and several major eruptions occurring in 1631, 1794, 1872, 1906 and in 1944 in the midst of the Italian campaign of World War II.

In the United States on March 27, 1980, Mount St. Helens Volcano in the Cascade Range, southwestern Washington, reawakened after more than a century of dormancy and provided a dramatic and tragic reminder that there are active volcanoes in the "lower 48" states as well as in Hawaii and Alaska. The catastrophic eruption of Mount St. Helens on May 18, 1980, and related mudflows and flooding caused significant loss of life (57 dead or missing) and property damage (over $1.2 billion). Mount St. Helens is expected to remain intermittently active for months or years, possibly even decades.

The word volcano comes from the little island of Vulcano in the Mediterranean Sea off Sicily. Centuries ago, the people living in this area believed that Vulcano was the chimney of the forge of Vulcan—the blacksmith of the Roman gods. They thought that the hot lava fragments and clouds of dust erupting from Vulcano came from Vulcan's forge as he beat out thunderbolts for Jupiter, king of the gods, and weapons for Mars, the god of war. In Polynesia the people attributed eruptive activity to the beautiful but wrathful Pele, Goddess of Volcanoes, whenever she was angry or spiteful. Today we know that volcanic eruptions are not supernatural but can be studied and interpreted by scientists.

The Nature of Volcanoes

Volcanoes are mountains but they are very different from other mountains. They are not formed by folding and crumpling but by uplift and erosion. Instead, volcanoes are built by the accumulation of their own eruptive products—lava, bombs (crusted over ash flows), and tephra (airborne ash and dust). A volcano is most commonly a conical hill or mountain built around a vent that connects with reservoirs of molten rock below the surface of the Earth. The term *volcano* also refers to the opening or vent through which the molten rock and associated gases are expelled.

Driven by buoyancy and gas pressure the molten rock, which is lighter than the surrounding solid rock, forces its way upward and may ultimately break though zones of weaknesses in the Earth's crust. If so, an eruption begins, and the molten rock may pour from the vent as non-explosive lava flows, or it may shoot violently into the air as dense clouds of lava fragments. Larger fragments fall back around the vent, and accumulations of fallback fragments may move downslope as ash flows under the force of gravity. Some of the finer ejected materials may be carried by the wind only to fall to the ground many miles away. The finest ash particles may be injected miles into the atmosphere and carried many times around the world by stratospheric winds before settling out.

Molten rock below the surface of the Earth that rises in volcanic vents is known as magma, but after it erupts from a volcano it is called lava. Originating many tens of miles beneath the ground, the ascending magma commonly contains some crystals, fragments of surrounding (unmelted) rocks, and dissolved gases, but it is primarily a liquid composed principally of oxygen, silicon, aluminum, iron, magnesium, calcium, sodium, potassium, titanium, and manganese. Magmas also contain many other chemical elements in trace quantities. Upon cooling, the liquid magma may precipitate crystals of various minerals until solidification is complete to form an igneous or magmatic rock....

Lava is red hot when it pours or blasts out of a vent but soon changes to dark red, gray, black, or some other color as it cools and solidifies. Very hot, gas-rich lava containing abundant iron and magnesium is fluid and flows like hot tar, whereas cooler, gas-poor lava high in silicon, sodium, and potassium flows sluggishly, like thick honey in some cases or in others like pasty, blocky masses.

All magmas contain dissolved gases, and as they rise to the surface to erupt, the confining pressures are reduced and the dissolved gases are liberated either quietly or explosively. If the lava is a thin fluid (not viscous), the gases may escape easily. But if the lava is thick and pasty (highly viscous), the gases will not move freely but will build up tremendous pressure, and ultimately escape with explosive violence. Gases in lava may be compared with the gas in a bottle of a carbonated soft drink. If you put your thumb over the top of the bottle and shake it vigorously, the gas separates from the drink and forms bubbles. When you remove your thumb abruptly, there is a miniature explosion of gas and liquid. The gases in lava behave in somewhat the same way. Their sudden expansion causes the terrible explosions that throw out great masses of solid rock as well as lava, dust, and ashes.

The violent separation of gas from lava may produce rock froth called pumice. Some of this froth is so light—because of the many gas bubbles—that it floats on water. In many eruptions, the froth is shattered explosively into small fragments that are hurled high into the air in the form of volcanic cinders (red or black), volcanic ash (commonly tan or gray), and volcanic dust.

Principal Types of Volcanoes

Geologists generally group volcanoes into four main kinds—cinder cones, composite volcanoes, shield volcanoes, and lava domes.

Cinder cones

Cinder cones are the simplest type of volcano. They are built from particles and blobs of congealed lava ejected from a single vent. As the gas-charged lava is blown violently into the air, it breaks into small fragments that solidify and fall as cinders around the vent to form a circular or oval cone. Most cinder cones have a bowl-shaped crater at the summit and rarely rise more than a thousand feet or so above their surroundings. Cinder cones are numerous in western North America as well as throughout other volcanic terrains of the world.

In 1943 a cinder cone started growing on a farm near the village of Parícutin in Mexico. Explosive eruptions caused by gas

Cinder cone volcanoes are created from the accumulation of millions of small pieces of lava, or "cinders."

rapidly expanding and escaping from molten lava formed cinders that fell back around the vent, building up the cone to a height of 1,200 feet. The last explosive eruption left a funnel-shaped crater at the top of the cone. After the excess gases had largely dissipated, the molten rock quietly poured out on the surrounding surface of the cone and moved downslope as lava flows. This order of events—eruption, formation of cone and crater, lava flow—is a common sequence in the formation of cinder cones.

During 9 years of activity, Parícutin built a prominent cone, covered about 100 square miles with ashes, and destroyed the town of San Juan. Geologists from many parts of the world studied Parícutin during its lifetime and learned a great deal about volcanism, its products, and the modification of a volcanic landform by erosion.

Composite volcanoes

Some of the Earth's grandest mountains are composite volcanoes—sometimes called stratovolcanoes. They are typically steep-sided, symmetrical cones of large dimension built of alternating layers of lava flows, volcanic ash, cinders, blocks, and bombs and may rise as much as 8,000 feet above their bases. Some of the most conspicuous and beautiful mountains in the world are composite volcanoes, including Mount Fuji in Japan, Mount Cotopaxi in Ecuador, Mount Shasta in California, Mount Hood in Oregon,

Geologists speculate that Oregon's Crater Lake was created by the eruption and collapse of a large volcano called Mount Mazama.

and Mount St. Helens and Mount Rainier in Washington.

Most composite volcanoes have a crater at the summit which contains a central vent or a clustered group of vents. Lavas either flow through breaks in the crater wall or issue from fissures on the flanks of the cone. Lava, solidified within the fissures, forms dikes that act as ribs which greatly strengthen the cone.

The essential feature of a composite volcano is a conduit system through which magma from a reservoir deep in the Earth's crust rises to the surface. The volcano is built up by the accumulation of material erupted through the conduit and increases in size as lava, cinders, ash, etc., are added to its slopes.

When a composite volcano becomes dormant, erosion begins to destroy the cone. As the cone is stripped away, the hardened magma filling the conduit (the volcanic plug) and fissures (the dikes) becomes exposed, and it too is slowly reduced by erosion. Finally, all that remains is the plug and dike complex projecting above the land surface—a telltale remnant of the vanished volcano.

An interesting variation of a composite volcano can be seen at Crater Lake in Oregon. From what geologists can interpret of its past, a high volcano—called Mount Mazama—probably similar in appearance to present-day Mount Rainier was once located at this spot. Following a series of tremendous explosions about 6,800 years ago, the volcano lost its top. Enormous volumes of volcanic ash and dust were expelled and swept down the slopes as ash flows and avalanches. These large-volume explosions rapidly drained the lava beneath the mountain and weakened the upper part. The top then collapsed to form a large depression, which later filled with water and is now completely occupied by beautiful Crater Lake. A last gasp of eruptions produced a small cinder cone, which rises above the water surface as Wizard Island near the rim of the lake. Depressions such as Crater Lake, formed by collapse of volcanoes, are known as calderas. They are usually large, steep-walled, basin-shaped depressions formed by the collapse of a large area over, and around, a volcanic vent or vents. Calderas range in form and size from roughly circular depressions 1 to 15 miles in diameter to huge elongated depressions as much as 60 miles long.

Shield volcanoes

Shield volcanoes, the third type of volcano, are built almost entirely of fluid lava flows. Flow after flow pours out in all directions from a central summit vent, or group of vents, building a broad, gently sloping cone of flat, domical shape, with a profile much like that of a warrior's shield. They are built up slowly by the accretion of thousands of highly fluid lava flows called basalt lava that spread widely over great distances, and then cool as thin,

The broad, gentle slopes of shield volcanoes are slowly built up by the eruption of fluid basalt lava.

gently dipping sheets. Lavas also commonly erupt from vents along fractures (rift zones) that develop on the flanks of the cone. Some of the largest volcanoes in the world are shield volcanoes. In northern California and Oregon, many shield volcanoes have diameters of 3 or 4 miles and heights of 1,500 to 2,000 feet. The Hawaiian Islands are composed of linear chains of these volcanoes including Kilauea and Mauna Loa on the island of Hawaii—two of the world's most active volcanoes. The floor of the ocean is more than 15,000 feet deep at the bases of the islands. As Mauna Loa, the largest of the shield volcanoes (and also the world's largest active volcano), projects 13,677 feet above sea level, its top is over 28,000 feet above the deep ocean floor.

In some eruptions, basaltic lava pours out quietly from long fissures instead of central vents and floods the surrounding countryside with lava flow upon lava flow, forming broad plateaus. Lava plateaus of this type can be seen in Iceland, southeastern Washington, eastern Oregon, and southern Idaho. Along the Snake River in Idaho, and the Columbia River in Washington and Oregon, these lava flows are beautifully exposed and measure more than a mile in total thickness.

Lava domes

Volcanic or lava domes are formed by relatively small, bulbous masses of lava too viscous to flow any great distance; consequently, on extrusion, the lava piles over and around its vent. A dome grows largely by expansion from within. As it grows its outer surface cools and hardens, then shatters, spilling loose fragments down its sides. Some domes form craggy knobs or spines over the volcanic vent, whereas others form short, steep-sided lava flows known as "coulees." Volcanic domes commonly occur within the craters or on the flanks of large composite volcanoes. The nearly circular Novarupta Dome that formed during the 1912 eruption of Katmai Volcano, Alaska, measures 800 feet across and 200 feet high. The internal structure of this dome—defined by layering of lava fanning upward and outward from the center—indicates that it grew largely by expansion from within. Mont Pelée in Martinique, Lesser Antilles, and Lassen Peak and Mono domes in California are examples of lava domes. An extremely destructive eruption accompanied the growth of a dome at Mont Pelée in 1902. The coastal town of St. Pierre, about 4 miles downslope to the south, was demolished and nearly 30,000 inhabitants were killed by an incandescent, high-velocity ash flow and associated hot gases and volcanic dust.

Only two men survived; one because he was in a poorly ventilated, dungeon-like jail cell and the other who somehow made his way safely through the burning city. . . .

Submarine Volcanoes

Submarine volcanoes and volcanic vents are common features on certain zones of the ocean floor. Some are active at the present time and, in shallow water, disclose their presence by blasting steam and rock-debris high above the surface of the sea. Many others lie at such great depths that the tremendous weight of the water above them results in high, confining pressure and prevents the formation and explosive release of steam and gases. Even very large, deep-water eruptions may not disturb the ocean surface.

The unlimited supply of water surrounding submarine volcanoes can cause them to behave differently from volcanoes on land. Violent, steam-blast eruptions take place when sea water pours into active shallow submarine vents. Lava, erupting onto a shallow sea floor or flowing into the sea from land, may cool so rapidly that it shatters into sand and rubble. The result is the production of huge amounts of fragmental volcanic debris. The famous "black sand" beaches of Hawaii were created virtually instantaneously by the violent interaction between hot lava and sea water. On the other hand, recent observations made from deep-diving submersibles have shown that some submarine eruptions produce flows and other volcanic structures remarkably similar to those formed on land. Recent studies have revealed the presence of spectacular, high temperature hydrothermal plumes and vents (called "smokers") along some parts of the mid-oceanic volcanic rift systems. However, to date, no direct observation has been made of a deep submarine eruption in progress.

During an explosive submarine eruption in the shallow open ocean, enormous piles of debris are built up around the active volcanic vent. Ocean currents rework the debris in shallow water, while other debris slumps from the upper part of the cone and flows into deep water along the sea floor. Fine debris and ash in the eruptive plume are scattered over a wide area in airborne clouds. Coarse debris in the same eruptive plume rains into the sea and settles on the flanks of the cone. Pumice from the eruption floats on the water and drifts with the ocean currents over a large area. . . .

Volcano Environments

There are more than 500 active volcanoes (those that have erupted at least once within recorded history) in the world—50 of which are in the United States (Hawaii, Alaska, Washington, Oregon, and California)—although many more are hidden under the seas. Most active volcanoes are strung like beads along, or near, the margins of the continents, and more than half encircle the Pacific Ocean as a "Ring of Fire."

Many volcanoes are in and around the Mediterranean Sea. Mount Etna in Sicily is the largest and highest of these mountains. Italy's Vesuvius is the only active volcano on the European mainland. Near the island of Vulcano, the volcano Stromboli has been in a state of nearly continuous, mild eruption since early Roman times. At night, sailors in the Mediterranean can see the glow from the fiery molten material that is hurled into the air.

Very appropriately, Stromboli has been called "the lighthouse of the Mediterranean."

Some volcanoes crown island areas lying near the continents, and others form chains of islands in the deep ocean basins. Volcanoes tend to cluster along narrow mountainous belts where folding and fracturing of the rocks provide channelways to the surface for the escape of magma. Significantly, major earthquakes also occur along these belts, indicating that volcanism and seismic activity are often closely related, responding to the same dynamic Earth forces.

In a typical "island-arc" environment, volcanoes lie along the crest of an arcuate, crustal ridge bounded on its convex side by a deep oceanic trench. The granite or granitelike layer of the continental crust extends beneath the ridge to the vicinity of the trench. Basaltic magmas, generated in the mantle beneath the ridge, rise along fractures through the granitic layer. These magmas commonly will be modified or changed in composition during passage through the granitic layer and erupt on the surface to form volcanoes built largely of nonbasaltic rocks.

In a typical "oceanic" environment, volcanoes are aligned along the crest of a broad ridge that marks an active fracture system in the oceanic crust. Basaltic magmas, generated in the upper mantle beneath the ridge, rise along fractures through the basaltic layer. Because the granitic crustal layer is absent, the magmas are not appreciably modified or changed in composition and they erupt on the surface to form basaltic volcanoes.

In the typical "continental" environment, volcanoes are located along unstable, mountainous belts that have thick roots of granite or granitelike rock. Magmas, generated near the base of the mountain root, rise slowly or intermittently along fractures in the crust. During passage through the granitic layer, magmas are commonly modified or changed in composition and erupt on the surface to form volcanoes constructed of nonbasaltic rocks.

Plate-Tectonics Theory

According to the now generally accepted "plate-tectonics" theory, scientists believe that the Earth's surface is broken into a number of shifting slabs or plates, which average about 50 miles in thickness. These plates move relative to one another above a hotter, deeper, more mobile zone at average rates as great as a few inches per year. Most of the world's active volcanoes are located

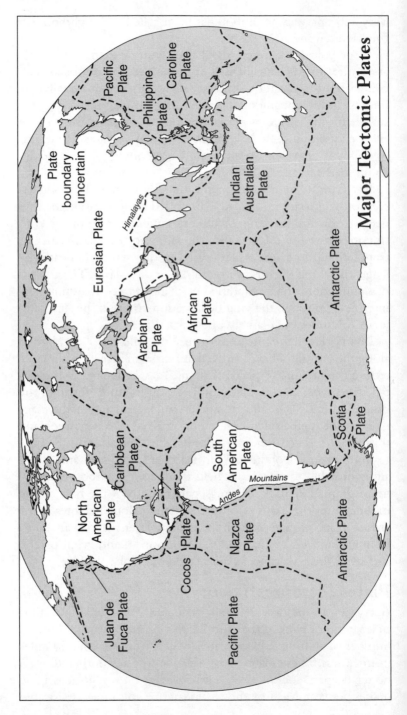

Major Tectonic Plates

along or near the boundaries between shifting plates and are called "plate-boundary" volcanoes. However, some active volcanoes are not associated with plate boundaries, and many of these so-called "intra-plate" volcanoes form roughly linear chains in the interior of some oceanic plates. The Hawaiian Islands provide perhaps the best example of an "intra-plate" volcanic chain, developed by the northwest-moving Pacific plate passing over an inferred "hot spot" that initiates the magma-generation and volcano formation process. The peripheral areas of the Pacific Ocean Basin, containing the boundaries of several plates, are dotted by many active volcanoes that form the so-called "Ring of Fire." The "Ring" provides excellent examples of "plate boundary" volcanoes, including Mount St. Helens. . . .

Extraterrestrial Volcanism

Volcanoes and volcanism are not restricted to the planet Earth. Manned and unmanned planetary explorations, beginning in the late 1960's, have furnished graphic evidence of past volcanism and its products on the Moon, Mars, Venus and other planetary bodies. Many pounds of volcanic rocks were collected by astronauts during the various Apollo lunar landing missions. Only a small fraction of these samples have been subjected to exhaustive study by scientists. The bulk of the material is stored under controlled-environment conditions at NASA's Lunar Receiving Laboratory in Houston, Tex., for future study by scientists.

From the 1976–1979 Viking mission, scientists have been able to study the volcanoes on Mars, and their studies are very revealing when compared with those of volcanoes on Earth. For example, Martian and Hawaiian volcanoes closely resemble each other in form. Both are shield volcanoes, have gently sloping flanks, large multiple collapse pits at their centers, and appear to be built of fluid lavas that have left numerous flow features on their flanks. The most obvious difference between the two is size. The Martian shields are enormous. They can grow to over 17 miles in height and more than 350 miles across, in contrast to a maximum height of about 6 miles and width of 74 miles for the Hawaiian shields.

Voyager-2 spacecraft images taken of Io, a moon of Jupiter, captured volcanoes in the actual process of eruption. The volcanic plumes shown on the image rise some 60 to 100 miles above the surface of the moon. Thus, active volcanism is taking place, at present, on at least one planetary body in addition to our Earth. . . .

Volcanoes and People

Volcanoes both harass and help mankind. As dramatically demonstrated by the catastrophic eruption of Mount St. Helens in May 1980 and of Pinatubo in June 1991, volcanoes can wreak havoc and devastation in the short term. However, it should be emphasized that the short-term hazards posed by volcanoes are balanced by benefits of volcanism and related processes over geologic time. Volcanic materials ultimately break down to form some of the most fertile soils on Earth, cultivation of which fostered and sustained civilizations. People use volcanic products as construction materials, as abrasive and cleaning agents, and as raw materials for many chemical and industrial uses. The internal heat associated with some young volcanic systems has been harnessed to produce geothermal energy. For example, the electrical energy generated from The Geysers geothermal field in northern California can meet the present power consumption of the city of San Francisco.

The challenge to scientists involved with volcano research is to mitigate the short-term adverse impacts of eruptions, so that society may continue to enjoy the long-term benefits of volcanism. They must continue to improve the capability for predicting eruptions and to provide decision makers and the general public with the best possible information on high-risk volcanoes for sound decisions on land-use planning and public safety. Geoscientists still do not fully understand how volcanoes really work, but considerable advances have been made in recent decades. An improved understanding of volcanic phenomena provides important clues to the Earth's past, present, and possibly its future.

The Volcano Watchers

By Tim Woody

Some of the most valuable work in volcanology is not glamorous. It consists of hiking for days through volcanic debris and collecting bag after bag of samples. This tedious and demanding work provides much of what is known about volcanic behavior around the world. In this article, Tim Woody, managing editor of Alaska *magazine, describes how this kind of work is done at the Alaska Volcano Observatory (AVO), the largest volcano research institute in the world.*

The AVO's job is to monitor volcanoes and alert people when it seems that one might erupt. The volcanologists at the AVO have many opportunities to study volcanoes as Alaska is home to 80 percent of U.S. volcanoes. Since written records have been kept beginning around 1760, more than forty active volcanoes have been identified in this area. In addition to monitoring volcanoes in Alaska and on the nearby Aleutian Islands, AVO volcanologists have been called to eruptions around the world for their help.

The Boeing 747 had flown overnight from Amsterdam and was hurtling through the dark sky of an early December morning in 1989. As they approached Anchorage, the 231 passengers and 14 crew members had no idea their airplane was about to be crippled and they would spend 13 minutes thinking they were about to die.

Ten hours earlier and 150 miles away, Redoubt Volcano had exploded, propelling a plume of volcanic ash more than 35,000 feet into the atmosphere. Unaware of the eruption, the KLM [airline] pilots flew directly into the cloud laden with silicon—which melted inside the turbine engines, clogging fuel nozzles and cooling holes.

The jet's four engines shut down almost immediately and the

Tim Woody, "The Volcano Watchers," *Alaska*, vol. 67, October 2001, pp. 46–51.

cabin filled with a fine, smokelike ash as the 747 plunged more than two miles toward snow-covered mountains. For eight terrifying minutes, the people aboard contemplated their doom. Then the pilots restarted two engines. Another five minutes passed before the other two engines sparked to life. The crew managed to land the jet in Anchorage, limiting the loss to dollars instead of lives. The toll: $80 million in repairs, including four totaled engines.

Each day, thousands of passengers and millions of dollars in cargo fly over more than 100 volcanoes at the northern edge of the Pacific Rim. The Alaska Peninsula and the Aleutian Islands contain more than 40 historically active volcanoes that erupt sporadically—often at night on remote, unpopulated islands where eruptions and the resulting hazards can go undetected.

Minding these volcanoes is the job of scientists at the Alaska Volcano Observatory, formed in 1988 to monitor the state's active volcanoes, notify the public of eruptions and hazards—in simpler terms, AVO gives people an idea of how and when a mountain might blow up so they can stay out of the way.

The Tough Work of Volcanology

Volcanologists Cynthia Gardner and Kathy Cashman bound over the remains of debris avalanches on the southern flank of Augustine Volcano, quickly stepping from boulder to boulder as I struggle to keep up. During the rare moments I'm within earshot, Gardner is usually humming or singing as we pick our way over the ankle- and knee-twisting rocks that have rolled down Augustine's steep slopes over the centuries.

This active volcano rises from the sea off the Alaska Peninsula only 174 miles southwest of Anchorage, casting its classic, conical silhouette across lower Cook Inlet. When conditions are right, steam rises from the summit as hot, moist air vents into the cool atmosphere. Augustine's 1986 eruption was the most recent of six over the past two centuries and the volcano is layered with deposits of ash, rock debris and pyroclastic flows—the mixtures of hot gases and fragmented lava that can move at more than 100 mph during an eruption.

I've come to this unpopulated island to observe volcano researchers in the field, and so far it's tough work. We're hiking several miles along a beach cluttered with volcanic boulders. Huge driftwood logs and lost fishing buoys are lodged well above the high tide line, testifying to the power of storms that occasionally pound this coast. We're headed to a seaside bluff where Gardner and Cashman will excavate ancient deposits of tephra, the material ejected during eruptions.

Equipped with notebooks, plastic bags, garden tools and military-surplus pack shovels, Gardner, from the Cascades Volcano Observatory in Vancouver, Washington, and Cashman, a professor at the University of Oregon, will spend most of the day hiking over boulders and loose sand—and sometimes through tangled, head-high vegetation. Once at their destination, they'll spend hours scrambling up crumbling bluffs and gathering samples that were deposited before humans began keeping track of Augustine's activity.

These women don't fit the Hollywood image of volcanologists. They're not buzzing around in helicopters trying to save a town from imminent destruction. They had the luxury of a helicopter for only the first day of this trip when the aircraft was used to drop gear at our camp and access coastal bluffs too difficult to reach on foot. During their week here they'll cover miles of the island's coast on foot and then pack up and hike for several hours to meet a floatplane at a lagoon on the west side of the island. The research they're doing is part of a project they started in 1998 to study the explosive behavior of Cook Inlet volcanoes. Gardner—who has studied Alaska volcanoes since the 1980s and has a long working relationship with AVO—summarizes their goal simply: to find out what Augustine is capable of in the future.

"The big question is, what's it going to do and when's it going to do it?" Cashman says.

Augustine lies within 186 miles of half the population of Alaska and has sent ash plumes as high as 39,000 feet, endangering aircraft and creating a nuisance when it falls on populated areas. Its eruptions also disrupt airline flight schedules, shipping and oil and gas operations. Ominously, its steep flanks are prone to large debris avalanches that could slam into Cook Inlet, setting off a tsunami that could strike much of the southern Kenai Peninsula. Imagine a 30-foot wall of water racing silently toward the Homer Spit during high tide. A worst-case scenario? Perhaps, but it's not the stuff of fiction. A tsunami that struck a low-lying area of New Guinea in 1998 killed thousands of people.

And Augustine is a strong candidate to be among the next few Alaska volcanoes to erupt.

The Alaska Volcano Observatory

The Alaska Volcano Observatory is an unusual amalgam of goverment agencies, combining elements of the U.S. Geological Survey, the Geophysical Institute of the University of Alaska Fairbanks and the Alaska Division of Geological and Geophysical Surveys. The observatory offices are in Anchorage—the primary source of public information during a crisis—and Fairbanks, where most seismic and satellite data are collected. "AVO is the flagship of the United States volcano hazards program," said John Dehn, a volcanologist in the Fairbanks office who specializes in remote sensing—using satellite imagery to detect volcanic activity and track ash clouds that can drift for thousands of miles. Dehn said AVO is the largest and most cost-efficient volcano observatory in the world.

Indeed, staff member Chris Nye said, AVO operates on a budget of about $5 million per year and "We feel like we're a bargain."

The observatory's long-term goal is to have seismic networks on all 41 of the Alaska volcanoes known to have erupted since record keeping began with the arrival of Russian sailors and fur trappers in 1760. Such monitoring poses a daunting task. Scientist-in-charge Tom Murray said installing seismic networks on some volcanoes could cost $1 million each, partly because of the high cost of chartering boats and helicopters to get to remote places.

But the cost of not monitoring them could be higher. Nye compares flying large passenger jets over Aleutian volcanoes to "essentially taking a small Midwestern city and putting it at risk three or four days a year," which is how frequently eruptions occur along Alaska's so-called "arc of fire."

Sitting in the sand beside a camp stove and looking across Kamishak Bay toward Mount Douglas on the Alaska Peninsula, Gardner jokes that by the end of my three-day stay I'll be weary of watching her and Cashman look at dirt and rocks. Life on the island isn't glamorous—freeze-dried food, rocks for chairs, bloody insect bites and cold rain that keeps us in our tents most of a day. When the weather is clear, a day's work means hiking and digging.

Gardner and Cashman are focusing on ancient deposits that indicate Augustine's early eruptions were larger and more explosive than more recent events. They want to know what to expect if the volcano returns to earlier patterns of behavior.

Although it's far from action packed, this type of methodical, laborious work provides much of what is known about the behavior of volcanoes around the world. Although they might never be able to accurately predict when volcanoes will erupt, scientists hope to at least understand how powerful eruptions might be. And each volcano can teach scientists lessons that might apply to similar volcanoes elsewhere.

"The problem with predictions and forecasts of eruptions is that the history of most volcanoes includes, at most, five or six observations (of eruptions)," Cashman says. "Except in Japan, where the written history goes back 1,000 years."

The AVO's Work

The AVO staff is made up of about 50 people from the agencies that comprise the observatory. They work in a variety of full-time and part-time roles, forming the equivalent of about 25 full-time positions. They monitor 22 volcanoes with seismic equipment that transmits data to telephone networks, through which it is instantly fed to the observatory. Although eruptions can occur suddenly, volcanoes typically send signals prior to dangerous activity. Magma isn't stealthy; as it rises toward the surface, it fractures rock and changes a mountain's shape, causing it to rumble and shake. Detecting such earthquakes helps scientists know when an eruption might be imminent—or already occurring.

Without instruments or witnesses to report activity, it is diffi-

Researchers install a seismic station that will be used to monitor and predict volcanic activity.

cult to know if a volcano is about to erupt. But once it blows its top, volcanologists can detect the erupted material by looking at infrared satellite images.

A new piece of software created at AVO scans satellite images 24 hours a day, Dehn said, and should increase safety by detecting ash clouds when scientists aren't on duty—such as overnight. When an eruption is detected, the system transmits a message to the scientist in charge via his pager so that he can alert other AVO staff members and they can get to work monitoring the event and issuing warnings. Such a system is especially important for volcanoes like those in the Aleutians, where there might be no witnesses to an eruption that could put an ash cloud in the path of a morning jetliner loaded with hundreds of people.

When a volcano is threatening to erupt, AVO staff members often work late to check for ash clouds in the wee hours of the morning. Still, surprises do occur.

"We really need to improve our information network out there," Dehn said of the Aleutians. When Mount Cleveland erupted February 19, 2001, its ash cloud rose to 35,000 feet. The eruption had been occurring for three hours before it was discovered at 9 A.M. as Dehn and other staff members arrived at work and began looking at satellite images.

When an eruption is detected, Murray notifies the Federal Aviation Administration, National Weather Service, Elmendorf Air Force Base and the Volcanic Ash Advisory Centers, through which the information is distributed worldwide.

Because most of Alaska's volcanoes do not directly threaten cities, AVO's focus is on keeping aircraft out of harm's way, but there are communities that rely on AVO to keep them informed.

Previous eruptions have made hazard assessments—in Alaska and abroad—a priority in the field of volcanology. It's a priority forged in painful lessons.

Volcanology Coming of Age

The closely observed eruption of Washington's Mount St. Helens in 1980 might have given rise to the modern science of volcanology, but scientist-in-charge Murray said a 1984 eruption in Colombia prompted its coming of age. AVO staff members and other volcanologists are sometimes called to eruptions around the world. Those called to Colombia recognized that lahars—swift, deadly flows of mud and debris down the side of an erupting volcano—threatened the village of Armero, but they believed the Colombian government would warn its citizens. The villagers were not told of the danger, and when the eruption occurred 23,000 people were entombed by mud. "We found out it was our responsibility," Murray said. "We have an obligation to do more than just the science part of it."

As part of that role, AVO prepares hazard assessments for specific Alaska volcanoes, giving governments and emergency services the information needed to plan for eruptions that might occur near populated areas. Two volcanoes—Redoubt and Mount Spurr—erupted in the 1990s within view of Anchorage, the state's largest population center.

Murray cited a 1996 Akutan Island scare as "one of the best hazard mitigations we've ever done." A swarm of small earthquakes on the volcanic island northeast of Unalaska had locals on edge. Many workers at a local seafood plant had experienced eruptions in Mexico and the Philippines, and they were ready to flee. Their employers were considering an evacuation. "These guys were nervous," Murray said.

Scientists from AVO set up seismic monitors, assessed the volcano and location of the island's population and determined that it did not pose a risk. John Power, a scientist who was at the Mount Pinatubo eruption in the Philippines in 1991, assured people there was no need to evacuate. The volcano later quieted down without erupting, ending what Murray considers a success story.

Another of AVO's long-term goals is to document a 10,000-year eruption history on each volcano in Alaska's arc of fire.

AVO scientists are also developing techniques for better understanding what happens inside a volcano before it erupts: How big is the magma chamber? How deep? Is the mountain inflating with rising magma, or deflating? How are the gases and chemicals in the magma behaving?

Alaska contains 80 percent of the nation's and 8 percent of the world's active volcanoes, including the full range of volcano types. Some volcanoes here pour out lava for months at a time. Others explode violently. Abundance and variety make the state a thriving laboratory for volcanologists.

"I think the big breakthroughs in volcanology are going to come from up here," Murray said, because scientists in Alaska can study all sorts of volcanoes and are guaranteed frequent eruptions.

Even if the research isn't the stuff of Hollywood blockbusters.

Inside a Volcano

By Donovan Webster

Volcanology sometimes requires scientists to get dangerously close to volcanoes. In recent years, more than thirty volcanologists have died pursuing the secrets of volcanoes. In the following selection, Donovan Webster describes an expedition to two active volcanic vents on Ambrym, one of eighty islands making up a country called Vanuatu located in the South Pacific. Sulfuric-acid rain eats away metal eyeglass frames, lava boils, and the earth rumbles not far from where the explorers camp. The first part of the expedition takes them through clouds of ashfall up the side of the crater of Marum or Niri Taten, meaning "mad pig," where they observe the crater's lava lake bubbling and spewing. The following day the explorers hike to the Benbow crater where the author and a photographer lower themselves by rope into the crater's pit for a better view. Webster captures the sense of awe and power evoked by these volcanoes.

Donovan Webster, former senior editor for Outside *magazine, has written for* Smithsonian, *the* New Yorker, *the* New York Times Magazine, *and* Wired *magazine.*

The volcano's summit is a dead zone, a cindered plain swirling with poisonous chlorine and sulfur gases, its air further thickened by nonstop siftings of new volcanic ash. No life can survive this environment for long. On the ash plain's edge, always threatening to make the island an aboveground hell, sit two active vents, Marum and Benbow, constantly shaking the earth and spewing globs of molten rock into the air. Yet across the black soil of the plain come all nine of us, a team of explorers, photographers, and a film crew, a volcanologist, and me. We have hacked through dense jungles on this island called Ambrym, one of some 80 islands making up the South Pacific nation of Vanuatu, and entered this inhospitable landscape to camp and explore for two weeks. We've tightroped up miles of eroded, inches-wide ridgeline—with deep canyons plummeting hundreds of feet on either side to totter at the lip of the volcanic pit of Benbow. The pit's malevolent red eye—obscured by gases and a balcony ledge of new volcanic rock—sits just a few hundred feet below.

"OK, your turn," Chris Heinlein shouts above the volcano's roar.

A sinewy and friendly German engineer, Heinlein hands me the expedition's climbing rope, which leads down, inside the volcano. Clipping the rope into a rappelling device on my belt—which helps control my descent—I step into the air above the pit.

A dozen feet of rope slips between my gloved fingers. I lower myself into the volcano. Acidic gas bites my nose and eyes. The sulfur dioxide is mixing with the day's spitting drizzle, creating a sulfuric-acid rain so strong it will eat the metal frames of my eyeglasses within days, turning them to crumbly rust. The breathing of Benbow's pit is deafening, like up-close jet engines mixed with a cosmic belch. Each new breath from the volcano heaves the air so violently my ears pop in the changing pressure—then the temperature momentarily soars. Somewhere not too far below, red-hot, pumpkin-size globs of ejected lava are flying through the air.

I let more rope slip. With each slide deeper inside, I can only wonder: Why would anyone do this? And what drives the guy on the rope below me—the German photographer and longtime volcano obsessive Carsten Peter—to do it again and again?

We have come to see Ambrym's volcano close up and to witness the lava lakes in these paired pits, which fulminate constantly but rarely erupt. Yet suspended hundreds of feet above lava up to 2200 degrees F that reaches toward the center of the Earth, I'm also discovering there's more. It is stupefyingly beautiful. The enormous noise. The deep, orangy red light from spattering lava. And those dark and brittle strands called Pele's hair: Filaments of lava that follow large blobs out of the pit, they cool quickly in the updraft and create six-inch-long, glassy threads that drift on the wind. It is like nowhere else on Earth.

The Allure of Volcanoes

Our first night on Ambrym we make camp in a beachside town called Port-Vato at the base of the 4,167-foot-high volcano. Shortly after sunrise the next morning, at the start of a demanding hike up the side of the volcano—walking a dry riverbed through thick jungle—I try to extract Peter's reasons for coming. As we crunch along the floor of black volcanic cinders, scrambling over shiny cliffs of cold lava that become waterfalls in the rainy season, Peter, 41, is grinning with excitement. Overhead dark silhouettes of large bats called flying foxes crease the morning sky like pterodactyls.

"I was 15 years old and on vacation in Italy with my parents. They took me to see Mount Etna," he says. "As soon as I saw it, I was drawn to the crater's edge. I was fascinated. My parents went back to the tour bus. They honked the horn for me to come—but I couldn't leave. I edged closer, seeing the smoke inside, imagining the boiling magma below. At that moment I became infected."

Since then Peter has traveled the world examining volcanoes. His trips have taken him to Iceland, Ethiopia, Indonesia, Hawaii, and beyond. "And of course," he says, "I have been back to Etna, my home volcano, many times."

Using single-rope descending and climbing techniques developed by cave explorers and adapted for volcanoes, Peter has been dropping into volcanoes now for nearly a decade. "The size and power of a volcano is like nothing else on Earth," he says. "You think you understand the Earth and its geology, but once you look down into a volcanic crater and see what's there, well, you realize you will never completely understand. It is that powerful. That big." He grins. "You'll find out what I mean, I think."

Camping on the Ash Plain

After a five-hour walk uphill I get my first glimpse of that power as the expedition emerges from the steep, heavily vegetated sides of the volcano's cone and onto the caldera. In the course of a few hundred yards the trail flattens out, and the palm trees and eight-foot-tall cane grasses that lushly lined the trail behind us become gnarled and dead, their life force snuffed by a world of swirling gas clouds and acid rain.

This is Ambrym's ash plain. Seven miles across, it's a severely eroded ash-and-lava cap hundreds of feet thick. Across the plain Benbow and Marum jut almost a thousand feet into the sky.

To protect ourselves from the harsh environment, our team quickly establishes a base camp near the caldera's edge. Shielded behind a low bluff separating the caldera from the jungle, the camp stretches through a grove of palms and tree-size ferns, the black soil dotted with purple orchids bobbing on long green stalks. For the remainder of this first afternoon we set up tents and create acid-rain-tight storage areas. The camp is a paradise perched on the edge of disaster. As night falls, we eat chicken soup fortified with cellophane noodles and plan tomorrow's exploration, the volcano rumbling regularly in the background as we talk.

After dinner we follow Carsten Peter to the edge of the ash plain and watch the vents light the gas clouds, wreathing each peak in ghostly red glows. "Look there," Peter says, pointing to a third red cauldron halfway up Marum's side. "That must be Niri Taten. Tomorrow we'll start there."

All night long the rumbling keeps awakening us. Just a few miles away lava boils and the Earth roars while each of us—lying quietly in a flimsy tent—anxiously dreams of those swirling red clouds. Tomorrow night at this time, I resolve as I drift off to sleep once again, one thing is certain: It will have been a day like none I've ever had.

A Small Mad Pig

In the morning, shortly after a sunrise breakfast, we strike out toward Niri Taten, several miles uphill. As we follow dry and eroded riverbeds toward the volcanic cones, a gentle rain falls.

"What does Niri Taten mean?" I ask our local guide, Jimmy.

"Niri Taten is a small pig," he replies. "A small mad pig. A crazy pig. A small pig that causes trouble to men."

Haraldur Sigurdsson, one of the world's premier volcanologists, walks alongside me in the dry riverbed, examining sheer cliff faces. He points out strata of tephra, a mixture of volcanic material. By examining these layers, volcanologists can tell a volcano's level of activity. Larger and coarser tephra far from a volcanic pit means a more powerful volcano, since heavier matter is thrown farther as more explosive energy is supplied.

It's true. The closer we hike to the craters, the more the character of the riverbed beneath our feet changes from silty black grit to charcoalsize stones—not unlike old-time furnace clinkers. "Each volcano has its own chemical fingerprint," Sigurdsson says. "Each volcano's mineral and elemental content is different because of the nature of the volcano itself: its rock and the shape of its vent. It helps volcanologists a lot in their study.

"Like the Tambora eruption of 1815 in Indonesia," Sigurdsson says. "We've found Tambora ash by its particular chemical signature almost everywhere on Earth. One of that magnitude happens about every thousand years." The Tambora explosion is said to have given off so much ash and sulfur dioxide—both of which blocked and reflected sunlight—that 1816 was a "year without a summer" across much of the world. There was crop-killing frost throughout the summer in New England. In northern Europe harvests were a disaster.

Suddenly, from two miles upwind behind us, Benbow gives a huge belch. We turn to look back. "Uh-oh," Sigurdsson says. "Ashfall on the way." Instead of the usual bluish white clouds of steam and gas, the plume issuing from the cone is heavy and black, trailing earthward in a dark curtain. Slowly it drifts our way on the heavy wind. Five minutes later the ashfall finds us, covering our rucksacks, clothing, faces, boots, and ponchos with a sandy grit the color of wet cocoa mix.

Under the ashfall we climb Marum, pressing forward through the dead volcanic soil for another hour. Each step takes us closer

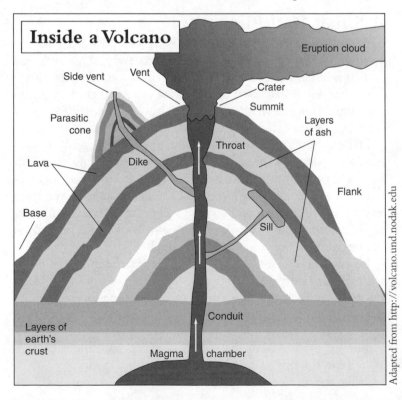

Inside a Volcano

Eruption cloud

Side vent · Vent · Crater · Summit

Parasitic cone

Layers of ash

Lava · Dike · Throat

Flank

Base

Sill

Conduit

Layers of earth's crust

Magma chamber

Adapted from http://volcano.und.nodak.edu

to Niri Taten, a crater that tunnels straight down into the basaltic rock like a massive, steaming worm burrow 200 yards across. As we approach, a rising wind and thick clouds of chlorine gas force us to pause and pull on safety helmets and industrial-style gas masks that cover our noses and mouths. Without them, between the flying bits of stone and grit carried on the 50-mile-an-hour winds and the thick clouds of gas roaring upward from the vent, time spent near the pit's lip would be painfully dangerous if not impossible.

Even with these protections the howling wind and gas often force us to shut our eyes and suspend breathing until the heaviest gas clouds pass. We lean against the high winds, brace at the crater's edge, and look inside.

Five hundred feet below, the vent's opening is obscured by rocky ledges. But if we can't see the lava itself, there is a consolation. Every inch of rocky surface inside the vent's cone is painted with color. Sunshine yellow sulfur coats some of the crater's sheer rock faces. Iron washes other sections of rock with flaming orange. Pastel green deposits of manganese glaze rock nearest the vent, like a carpet of immortal moss. Other patches of stone have been bleached white by chlorine and fluorine gases pouring from the vent.

Besides the wind and dangerous concentration of gases, the edges of Niri Taten are too crumbly to allow safe descent. Anyone climbing down a rope inside the crater could be dislodging loose boulders, some the size of cars, that could crash on anyone below. Carsten Peter pulls out his camera and long lens—whose coating immediately becomes corroded in the noxious air. The howling gusts twice knock expedition members to the ground.

Marum's Lava Lake

After an hour it's decided that we should examine the Marum crater itself. "We can get two volcanoes in one day!" Carsten Peter says with glee. Our helmeted heads tucked down, we continue breathing scuba-diver slow into our masks for maximum benefit, and we push on.

The walk to Marum's opening isn't far, but what it lacks in distance it makes up for in danger. No matter which route you choose, you have to traverse the mountain's steep slopes, many of which are gouged with deep, unclimbable erosion gullies. We decide to cross where the gullies are smallest: along Niri Taten's knife-edged lip, within a foot of a sheer drop into the crater. We step gingerly where the slope looks most reliable, but our footing remains dangerously slick. The slope's top layer is crumbly tephra, sometimes as big as charcoal briquettes. Making things more difficult, we've moved downwind of Niri Taten. All around us clouds of sulfur dioxide, chlorine, and fluorine gases swirl so thick they sometimes obscure our vision and force us to stop and bury our gas-masked faces inside our arms for extra protection. It's a slog. Minutes stretch into an hour. Every step could be our last. Finally we reach the summit of the crater's edge and begin down its other side. Protected by the lip behind us, the environment changes. Sunshine blankets the tilting black ash, and the cold gales calm into balmy breezes.

Two expedition members, Franck Tessier and Irene Margaritis, hustle downslope with me toward Marum. As we approach its lip, the 39-year-old Tessier—a genial and easygoing French biologist with impressive rope and rock-climbing skills honed by years of adventures like this one—rips off his gas mask and begins to hoot with pleasure.

I know why. Ahead of us Marum's volcanic pit stretches as open and clear as a visionary's painting. In the pit, three step-down ledges—each deeper and wider than the one uphill of it—are marbled with layers of black ash and pale, bleached basaltic andesite. The layers of lava inside the vent form as a crust over a cooling lava lake that gets blown out like a massive champagne cork when volcanic activity resumes. Small wall vents called fumaroles—created where heated groundwater and escaping volcanic gas reach the surface—let off steady plumes of steam. Inside Marum's crater it looks as if the world is being born.

And there, in the bottom of the third and largest pit—some 1,200 feet below—sits the lava lake. Its fury pushes lava through three skylight holes in a roof that partly covers the lake like a canopy. Bright orange-and-red spatters fly unpredictably from the circular opening of the largest skylight, a hole perhaps 50 yards across.

Lava is three times as dense as water. Despite its up to 2200 degrees F heat, lava moves, burbles, and flies through the air with the consistency of syrup. Every few minutes huge molten blobs seem to soar in slow motion. A second or two later a noise from beneath the earth—a rumbling booooom—fills the pit and rolls across the sculpted ash plain beyond. It's mesmerizing: lava sloshing back and forth, bubbles emerging and popping like a thick stew. As we survey Marum's lip and crater, I can't take my eyes off the lava. Suddenly I understand Peter's obsession. As evening cloaks the pit's deepest recesses in shade, the lava lake and explosive bubbles glow more seductively. The spatterings glisten like enormous, otherworldly fireworks as they sail through the shadowed air.

Inside Benbow's Pit

Dangling inside Benbow's crater the following afternoon, I have time to reflect. This morning we followed the narrow ridge to Benbow's pit—which was firm enough to climb down. We fixed our rope, ate lunch in a spitting acid rain, and began our descent into the volcano.

Now, on the rope below me, Carsten Peter works his way deeper inside the crater. I let more rope slide through my hands, easing myself deeper as well.

With each drop the air shakes more violently; the clouds of poison gas grow thicker.

Waves of pressurized air rumble past me.

Grasping the rope tightly, I halt my descent at the edge of an overhung cliff and stare deeper inside. The lava lake waits below, ejecting orange bombs and smaller drops. Then, in a heartbeat, a wall of thick clouds blows between me and the pit, enveloping everything around me in a world of gray. In the shuddering air and disorienting noise, gravity, direction, and time seem to fade away. There is only the volcano, its existence a direct result of two tectonic plates colliding below me. Benbow roars again. The earth shakes.

In this moment I know I've gotten close enough to the fire at the center of the Earth. At that same second the clouds part and Benbow reappears. Fumaroles smoke, and steam swirls from the pit's walls. The Technicolor wash swarms around me like a kaleidoscope. Below, Carsten Peter hits the end of the rope just above Benbow's explosive vent. He pulls a camera from his bag and lifts it to his eye.

Torrential rains will frustrate another attempt to explore Benbow. Then dissension breaks out among some of the expedition's porters who helped carry gear up the volcano's steep cone, and it becomes clear that the team will have to leave Ambrym as soon as possible. In a last-ditch, 18-hour marathon, team members drop 1,200 feet into Marum and photograph its lava lake nonstop. They emerge from the crater and find a fractious camp. Jimmy cannot persuade the disgruntled porters to bend, and the tension escalates. With a satchel full of photographs Carsten Peter finally agrees to abandon the volcano—even as he vows to return.

The Exciting Career of Volcanology

By Michael Kernan

*In the following selection, reporter Michael Kernan describes the re-
search career of Richard Fiske, an internationally recognized volcanologist. Among other posts, Fiske has been director of the Smithsonian's National Museum of Natural History and chief of the Office of
Geochemistry and Geophysics, where he was in charge of hundreds of
scientists. He has a passion for his work that is contagious. His fascina-
tion for the little-known world of underwater volcanoes has led him to
spend time forty-five hundred feet below sea level in the dark and
cramped quarters of a tiny Japanese submersible. He is considered by
some to be the world's expert on these volcanoes.*

*Fiske became interested in volcanoes at age ten. Later, while doing
graduate work with two of his favorite professors, he had the opportunity
to study Mount Rainier, a volcano in the Cascades that is expected to
erupt in the not-too-distant future. The field of volcanology can be dangerous, and Fiske talks about his colleagues and friends that have been
killed and about some of his own close calls. Nevertheless, his enthusiasm
for his work has not waned. Besides being featured on the PBS Mac-
Neil-Lehrer News Hour and on National Public Radio, Fiske helped
create a full-production motion picture,* Inside Hawaiian Volcanoes,
*with famed cinematographer Maurice Krafft and narrator Roger Mudd.
He condensed the film into a video with a teacher's guide and distributed
it nationally and internationally. Fiske thinks of himself as a volcano
"junkie," lucky to have made an exciting career of what he loves.*

Michael Kernan, a reporter for the Washington Post, *has written articles for* Life, Saturday Evening Post, Reader's Digest, *and* Ladies
Home Journal.

R ichard Fiske is internationally known for his work with volcanoes; in certain circles in Japan he is a legend. For several years this Smithsonian geologist has been studying undersea explosive volcanoes with Japanese marine scientists. Japan is intensely concerned with the phenomenon, since it depends so much on the sea. There is also the possibility that an undersea volcano might erupt catastrophically, though that hasn't happened for thousands of years. To say the least, the tidal waves caused by such an eruption would be devastating for this island nation.

For much of this decade, Fiske has been at it, going down in a tiny Japanese submersible to explore recently active submarine craters. It's dark down there at 4,500 feet. The only lights are the sub's own beacons. The only sounds are the whirring of the video cameras and the rattle of things from the ocean floor being grabbed up by the sub's metal arm and dropped into a basket on its front.

"I've set two records," Fiske told me. "I was the tallest person ever to go into the submersible, and the oldest." He is 6 feet 1 and a bit, which is, of course, a lot more notable in Japan than here, and when he last went down, a year ago, he was 64, which is notable only when you understand that retirement at 60 is enforced in most Japanese universities. "I had to get in there like a wet noodle and scrunch up," Fiske remarked. "But I'd do it again in a minute."

All right, so why undersea volcanoes? There are plenty in the open air. "Japan is one of the best places to study explosive submarine volcanoes. For some time people didn't recognize that they even existed. We didn't know they were there unless they happened to erupt." For Fiske, and the Japanese, too, the most fascinating part is the fabulous deposits of metal-bearing minerals left by these eruptions. "We're talking about five or ten cubic miles of debris. I mean, these eruptions can be big."

Many ore deposits are formed from submarine volcanoes. The point is, on land vast amounts of valuable minerals are ejected by volcanic eruptions, only to be carried away by the wind. "But in an undersea eruption, these emanations hit cold seawater and are precipitated. They end up as mineral deposits on the ocean floor." And that also has the attention of the Japanese, although there are no immediate plans to mine the deposits.

"Volcanoes attract attention," Fiske says. "They show clearly that this earth of ours is alive, and not just sitting there like the everlasting hills. Volcanoes grow and do all kinds of interesting things."

Interesting: that would be the word for Kilauea, the Hawaiian volcano that has been erupting continually since January 1983. Fiske spends a lot of time in Hawaii; he has been going there for 30 years. "We rappel deep into cracks looking for layers of ash between lava flows. They are an indicator of how Kilauea has been breaking apart over the last 1,000 years.

"As we speak, lava is pouring into the ocean, building new land. What I'm concentrating on is the fact that the volcano's magma is actually pushing its south flank into the sea."

The area being pushed aside at a rate of some 10 to 15 centimeters a year—geologically comparable to the speed of light—forms a gigantic rectangle, 50 miles across and three miles thick. "We're studying how the volcano is doing this and what the evidence is for its rates of motion in the past. The south flank of Kilauea, as it detaches, leaves cracks and faults, and we study these."

If this entire south side of the mountain ever slides off into the Pacific, "it would be truly apocalyptic. There might be tidal waves hundreds of feet high; every coastal community in Hawaii would be at risk."

Something like it has happened already: hundreds of thousands of years ago the eastern half of Oahu broke off and parts of it slid 200 miles across the ocean floor. It must have been a disaster of biblical proportions.

Risks of Being a Volcanologist

Fiske has never seen a volcano aborning, like Paricutín, which in 1943 grew out of a Mexican farmer's cornfield much to the farmer's irritation. But he has seen a lot of sudden activity. He has been lucky, too. "I was in a helicopter crash in the eastern Caribbean, which I regard as a close call. The helicopter was destroyed. I was so thoroughly strapped in that I wouldn't have made it out if the fuel tank had exploded. Other than that I've never been endangered by working on volcanoes." A number of Fiske's colleagues and friends have been killed, including the famous French husband-wife team of volcano cinematographers, Maurice and Katia Krafft, who traveled the world filming eruptions. "I met them 30 years ago and got to know them very well. They were good friends of the Smithsonian's Global Volcanism

Program and were also good personal friends. Wonderful people.
Very gregarious. They were serious volcanologists and knew the
dangers—they just became a little too careless. In 1991 while
filming a Japanese volcano called Unzen, they and 40 others were
caught in an avalanche of ash and hot gas. Gas masks or fire suits
wouldn't have protected them. They were just too close at the
wrong time. It would be an awful death—being baked in the extreme heat. Generally, you can move out of the way in time. Lava
flows slowly. But an explosion or an avalanche of hot gas can en-
gulf you in an instant."

When we met, Fiske was mourning the tragedy on Montserrat in the Caribbean. "The southern two-thirds of the island is
just devastated, everybody's left, 20 or so people killed and no end
in sight. This can go on for two or three years. It's terrible: people
have lost their property, their houses; people are still paying mort-
gages on property covered by ash." Everything was covered by
inches and feet of gray mud, the result of dozens of pyroclastic
flows. Buildings were buried to the rooftops.

Getting into Volcanoes

Fiske got into his work by luck: he had an uncle who was a geologist for an oil company in Venezuela, and at age 10 he decided
to do the same. "As a kid growing up in Baltimore it seemed
really romantic." For a while he had summer jobs for oil firms.
Petroleum geologists lead fairly eventful lives. Lots of travel.

But there was more than that. While getting his doctorate at
Johns Hopkins, Fiske joined forces with two of his favorite professors studying Mount Rainier. Like Mount St. Helens, Rainier
will doubtless erupt in the future and could send an enormous
river of mud and debris, called a "lahar," straight down upon
densely populated areas. From then on, for Dick Fiske, it was volcanoes. . . .

"I do like the fieldwork best, though," he said. "I sometimes
take my family to Hawaii on vacations, but when I'm there to
work there's little time for anything else. Once you get started in
volcanoes you become a junkie, I guess. The earth is always
changing, and you're trying to outfox it, understand its past activity and predict what it's likely to do in the future. Volcanoes
are exciting, and for me they have made for a wonderfully ful-
filling career."

Volcanic Disasters

Vesuvius: Monarch of the Mediterranean

By Kent H. Wilcoxson

Mount Vesuvius in Italy is the most famous volcano in history. Some people equate the name Vesuvius with volcano. Although it has erupted many times, the most famous eruption happened in A.D. 79. This eruption buried the cities of Pompeii and Herculaneum, and volcanologists believe that sixteen thousand people perished in the disaster. The following excerpt from Kent H. Wilcoxson's book Chains of Fire: The Story of Volcanoes *gives an account of that eruption, including some passages by Pliny the Younger, whose description of that event has been frequently cited. Pliny's first letter describes the beginning of the volcanic activity as well as the death of Pliny's uncle, Pliny the Elder, a Roman admiral. The second letter describes the flight of Pliny and his mother from Pompeii.*

The excavation of Pompeii is still incomplete, and scientists continue to study and observe Vesuvius.

I t was, perhaps, the two letters from Pliny the Younger to the historian Tacitus which first directed the western world's attention to the devastating power of volcanoes in general and of Vesuvius in particular. At the time, however, science had not yet come of age, and, although Stromboli and Etna were familiar to the ancients, such volcanic eruptions were attributed to the activities of the gods and to "pent-up" winds, rushing through subterranean passageways, carrying up fire from burning seams of coal and sulphur.

The volcanic origins of Vesuvius, not known to be eruptive,

had been hinted at by only a few of the more farsighted observers of natural history. The geographer Strabo wrote of the mountain:

> About these places rises Vesuvius, well cultivated and inhabited all round, except at its top, which is for the most part level, and entirely barren, ashey to the view, displaying cavernous hollows in the cineritious rocks, which look as if they had been eaten in fire; so that we may suppose this spot to have been a volcano formerly, with burning craters, now extinguished for want of fuel.

Pliny's First Letter

Then came the well-known cataclysm of A.D. 79, when Vesuvius destroyed Pompeii. The elder Pliny, a Roman admiral, lost his life while attempting to rescue and evacuate some of his friends from danger and at the same time satisfy his naturalist's curiosity about the nature of the outbreak. In his honor, paroxysmal eruptions of the Vesuvian cycle are often referred to as *Plinian*.

His nephew, known as Pliny the Younger, was asked by the leading historian, Tacitus, to supply some details of the catastrophe. His stirring though incomplete account of the cataclysm, and the consequent burial of Pompeii under the volcanic ejecta, took the form of two letters. The text of the first reads in part as follows:

> You ask me for an account of my uncle's death, so as to be able to hand down a more accurate report of it to posterity; thanks for this; I know that as related by you his death will immortalize his memory.
>
> He was at Misenum where he was commanding the fleet. On the ninth day before the Kalends of September [August 24], at about the seventh hour, my mother informed him that a cloud of extraordinary dimensions had been seen. He took his sun-bath, then a cold water one and, after eating lying down, he set to work. Having called for his sandals, he climbed to a spot from which he had a better view of this remarkable phenomenon. The cloud was rising; onlookers could not tell, at a distance, from which mountain; later it was known to be Vesuvius. More than any other tree, the

pine gives an idea of the shape and appearance of that
cloud. Projected into the air like an immense tree-
trunk, as it were, it opened out into branches. I imagine that taken up by a sudden gust which then died
down and left it, or defeated by its own weight, it scattered widely, now white, now dark and flecked, according to whether it bore along earth or ashes.

A grandiose spectacle, worth a scientist's while to study
at closer quarters. My uncle had one of his smaller craft
got into trim and gave me the option of accompanying him. I answered that I would rather work; as it happened, he himself had given me something to anno-
tate. As he was leaving the house, he received a note
from Rectina, Caesius Bassus' wife, who, alarmed at the
nearness of danger (for her villa was at the foot of the
mountain and escape was possible only by ship), begged
him to save her in this extremity. He changed his mind,
and what he had begun for the love of science, he now
continued out of humanity. Having the quadriremes
lowered, he took ship to rescue Rectina and many others as well, for this pleasant coast was extremely populous. He hurried to the places from which others were
fleeing; he steered eastwards, directing course straight
into danger, and was so immune from fear that he dic-
tated or noted down all the changing phenomena that
he observed.

Already ashes were falling on the boat, hotter and more
thickly as it drew nearer; and also pumice-stones, black,
ashen flints shattered by the fire. The sea, driven back,
was no longer deep enough; rubble from the mountain
made the shore inaccessible. Momentarily my uncle
considered turning back, and his pilot urged him to do
so. But "Fortune favours the brave" he said then.
"Make for Pomponianus."

Pomponianus lived in Stabiae, a town isolated by a cove
where little by little the sea is thrusting into the curve
of the shoreline. Here the danger was not yet immediate, but terrible all the same, and near at hand, what
with its inexorable advance. Pomponianus, resolved to
set sail as soon as the opposing wind had died down,

had loaded all his movables on to the ships. Favored by
this same wind, my uncle got to him, found him over-
wrought, but embraced and cheered and encouraged
him, and to reassure him with his own confidence went
to bathe. After that he sat down at the table and ate
with gusto or—which would be no less admirable—
with the appearance of gusto.

However, on several points of Mt. Vesuvius could be
seen shining great flames, huge conflagrations, their
brilliance and clarity enhanced by the night. And my
uncle, to allay his companions' fears, told them that this
was only country houses, abandoned by terrified rus-
tics to the consuming fires, and empty. Then he lay
down and slept a proper sleep, for those who stayed
near the door heard the sound of his breathing. Mean-
while the court on which his room abutted was filling
up with ashes and pumice-stones, heaped to such a
level that if he had waited any longer it would have
been impossible to get out. Wakened, he went to rejoin
Pomponianus and the others, who had not been rest-
ing. They took council. Should they stay in the house
or range into the country? Houses, shaken by frequent
and prolonged tremors, and as if torn from their foun-
dations, tilted to the right, to the left, then resumed
their original positions. In the open, there was the danger of the falling pumice stones, though these were
charred and light. Between these two hazards, it was the
latter that they chose, my uncle giving in to the best
policy, his companions replacing one fear with another.
With towels they fixed pillows to their heads, as protection against the falling stones.

Elsewhere the dawn had come but here it was night,
the blackest and thickest of nights, though counteracted
by numerous torches and lights of every kind. They
went back to the shore, to see from nearer at hand if
the sea would allow of an attempt; it was still tumul-
tuous and adverse.

Then my uncle laid himself down on a cloth spread
out for him and twice called for cold water and drank
it. Presently the flames, and the sulphurous odour

heralding their approach, put everyone to flight, forc-
ing my uncle to get up. Leaning on two young slaves,
he rose and immediately fell down dead. I suppose that
the thick smoke must have impeded his breathing and
closed his respiratory passages, which in his case were
naturally weak and narrow, often making him winded.
When light returned, three days after my uncle had
seen it for the last time, his body was found intact without any injury, not in the least disarrayed; he looked
more like a man asleep than a dead one.

Pliny's Second Letter

The second narrative describes the flight of Pliny and his mother
from Pompeii during the heavy fall of ash and pumice, which
was rapidly accumulating. This letter describes conditions as they
were at the time when the fallout of ejecta gave the choice of
flight or being buried alive.

It was the first hour of the day, however the light appeared to us faint and uncertain. The buildings around
us were so unsettled that, in this place which was open
to be sure, but narrow, the collapse of walls, seeming a
certainty, became a great menace. We decided to get out
of the town. The panic-stricken crowds followed us,
obeying that fear which makes it seem prudent to go
the way of others; in a long, close tide they harassed and
jostled us. Once clear of the houses, we stopped, and
there we encountered fresh prodigies and terrors. The
chariots we had taken there, though on level ground,
knocked about in every direction and even with stones
could not be kept steady. The sea appeared to have
shrunk into itself, as if pushed back by the tremors of
the earth. At all events, the banks had widened, and
many sea creatures were beached on the sand. In the
other direction gaped a horrible black cloud torn by
sudden bursts of fire in snake-like flashes, revealing
elongated flames similar to lightning, but larger.

And now came the ashes, though as yet sparsely. I
turned around. Ominous behind us, a thick smoke
spreading over the earth like a flood followed us. "Let
us get into the fields while we can still see the way," I

told my mother, for fear of being crushed by the mob around us in the road in the midst of this darkness. We scarcely agreed on this when we were enveloped in night, not a moonless night or one dimmed by cloud, but the darkness of a sealed room without lights. Only the shrill cries of women, the wailing of children, the shouting of men were to be heard. Some were calling to their parents, others to their children, others to their wives, knowing one another only by voice. Some wept for themselves, others for their relations. There were those who, in the very fear of death, invoked it. Many lifted up their hands to the gods, but a great number believed that there were no more gods and that this night was the world's last, eternal one. Some, with their false or illusory terrors, added to the real danger: "In Misenum," they would say, "such and such a building has collapsed, and some other is in flames." This might not be true, but it was believed.

A certain clearing appeared to us not as daylight but as a sign of approaching fire. It left off some distance from us however. Once more, darkness and the ashes, thick and heavy. From time to time we had to get up and shake them off for fear of being actually buried and crushed under their weight. I can boast of not having uttered a single lamentation, not a single word that might have been a sign of weakness, in so great a danger. I imagined that one with all, and all with one, were going to perish—a wretched but strong consolation in my dying. But this darkness lightened and then, like smoke or a cloud, dissolved away. Finally a genuine daylight came, the sun even shone, but pallidly, as in an eclipse. And there before our still-stricken gaze, everything appeared changed and covered, as by an abundant snowfall, with a thick layer of ashes.

The Destruction of Pompeii

The destruction of Pompeii was rapid and thorough. In less than two days, the city, five miles southeast of the volcano, was completely covered by the fallout of pyroclastic materials. Herculaneum, lying at the foot of Vesuvius, was also lost, both figuratively and literally. After the destruction had occurred, the exact sites of the buried cities were inexplicably forgotten over the centuries. There was some speculation as to their whereabouts, but nobody took the time or trouble to excavate to find their exact locations, and they lay buried for fifteen centuries. . . .

At the time of the devastation, Pompeii was a thriving commercial town and resort playground for the rich. The population was probably in the neighborhood of twenty thousand, of whom it is estimated that three-fourths perished in the holocaust.

When ashes and pumice began to fall, those who were wise evacuated immediately. It seems obvious that Pliny and his mother left the city before the fallout had accumulated to any great depth, but at a time when its intensity was becoming so great that anyone who attempted to leave much thereafter probably did not get away. Soon the fallout enveloped the town and obscured vision to such a degree that those who tried to flee must have groped and stumbled about blindly in the inky shroud of blackness under which they were to die. Those who hesitated in an effort to save valued possessions were lost. Many of the skeletons later found in the exhumed city were still clutching bags of gold and silver coins or jewels, which literally cost their owners their lives. . . .

Herculaneum Is Buried by Mud Slides

The town of Herculaneum was destroyed at essentially the same time as Pompeii, but the mode of destruction was entirely different. Herculaneum did not suffer from the tremendous fallout of pyroclastics and wind-borne gases which quickly plagued and overwhelmed Pompeii, simply because it was located on the upwind side of Vesuvius.

Though the sequence of events is not known exactly, it seems evident that large amounts of rain, which usually accompany eruptions of such magnitude, made a muddy slosh of the pyroclastics that fell and collected on the slopes of the volcano. When this mass had attained sufficient moisture, gigantic quantities of mud began to slide down its flanks.

Herculaneum lay immediately in the path of one of these mud slides and was submerged by it to a depth of sixty feet. Fortunately for those living there, the mud probably moved relatively slowly, allowing time for an almost complete evacuation of the town, as evidenced by the lack of skeletons in the areas excavated. . . .

Vesuvius's Past

Vesuvius, from the earliest of recorded times until A.D. 79, had remained deceptively dormant. Perhaps the first real indication of the activity which was forthcoming occurred in A.D. 63. In February of that year several severe earth shocks did damage in the vicinity of the Bay of Naples. Much of Pompeii was destroyed and had to be rebuilt. The quakes continued intermittently between A.D. 63 and 79, but no one guessed that they might precede an even greater calamity. Immediately before the outbreak, the last in a series of tremors rocked Pompeii. It was perhaps the release signified by this vibration which unleashed the fury of Vesuvius. . . .

Vesuvius has always dominated the volcanoes of the Mediterranean for a number of reasons. It has produced many spectacular eruptions in its long history, and these have been observed in detail by qualified scientists. It is perhaps the favorite volcano for study because it is small (approximately 3,800 feet), accessible, and approachable under almost all conditions, and because it has exhibited a larger variety of volcanic phenomena than any other single volcano in the world. If one, and only one, volcano had to be picked for a study of volcanology, Vesuvius would be the very best choice.

Proto-Krakatoa: The Unnamed Supervolcano

By David Keys

After four years of travel to more than one thousand archaeological sites in sixty countries, studying tree rings, polar ice cores, and ancient texts, archaeological journalist David Keys believes he has found enough substantial evidence to prove that one of the greatest natural disasters in world history happened in A.D. 535. His book, Catastrophe, *is a scientific reconstruction of the eruption of an Indonesian supervolcano, which he calls "proto-Krakatoa" because it was the site of the famous later eruption of Krakatau in 1883. According to his evidence, this gigantic eruption blocked the light and heat of the sun for eighteen months, crops failed in Asia and the Middle East as the weather radically changed, a bubonic plague infected Africa as the rodent population increased due to the weather change, and entire populations died in Europe. In this excerpt, Keys reconstructs the eruption and its disastrous impact on the climate of the entire planet and on the course of human history.*

David Keys is an archaeology correspondent for the Independent, *a London daily newspaper, and is an archaeology consultant for television.*

Reconstructing the immediate sequence of events associated with a volcanic eruption that occurred fifteen hundred years ago is a daunting task—but not an impossible one. Using historical, tree-ring, ice-core, and other data, it is possible to compare the event and its climatic consequences with more recent eruptions of known size and effect.

Using the quasi-historical account in the Javanese *Book of Ancient Kings*, it is possible, assuming the account to be at least part genuine, to gain an insight into specific aspects of the eruption itself. And using geological and volcanological knowledge of the area and records of more recent large eruptions, it is possible to reconstruct what probably happened.

Between 530 and 535, there would almost certainly have been a long series of earthquakes in what is now western Java, southern Sumatra, and the neighboring seas. These earthquakes and accompanying seismically triggered tidal waves may well have seriously disrupted life in the region. Typically, volcanic eruptions are preceded by increasingly frequent and violent tremors. Often the larger the eruption, the longer the seismic run-up to it will be.

In the case of the 530s catastrophe, the run-up to the eruption may even have included several earthquakes of level 6 on the Richter scale. Throughout the second half of 534, earthquakes would have struck the region at the rate of one or two per day. In the weeks immediately before the eruption, the rate would have accelerated to a peak of fifty quakes per hour in the final twenty-four hours, mainly in the range of 1 to 3 on the Richter scale.

Although it is a controversial proposal, it is geologically possible that Sumatra and Java were one island prior to the 535 supereruption—exactly as the Javanese *Book of Ancient Kings* describes. The 535 eruption would therefore have burst forth from a volcanic mountain located on fairly low-lying ground where the shallow Sunda Straits between Java and Sumatra are today. For several years, a huge mass of molten magma would have been moving closer and closer to the surface—probably at the rate of up to thirty feet per month. This would have caused the land surface above to bulge upward into a low dome, increasing in height at up to three feet per year over perhaps a five-year period.

The First and Second Phase

Then suddenly the pressure of the magma, two or three miles below the ground, would have proved too great; a crack would have opened up, and the first phase of the eruption would have started. A vast cloud of ash would have billowed forth, followed by a column of red-hot magma that would have shot out of the mountain like a fountain. A week or two later, as the magma came yet nearer to the surface, one of the earthquakes accompanying the eruption probably fractured the rock above the magma chamber, allowing the sea to rush into the wide tubes through which the magma was rising from the chamber to the surface.

The second phase began with a vast explosive event that shot even larger quantities of molten magma into the air at up to 1,500 miles per hour, reaching heights of perhaps thirty miles. The sound from this explosion would have broken the eardrums of most humans and animals living within a fifteen-mile radius.

The shock wave from the explosion would have moved outward at 750–1,500 miles per hour, devastating everything in its path for up to twenty miles. Houses, bridges, temples, and every single tree would have been leveled like so many matchsticks. And within an estimated ten-mile radius there would also have been massive fire damage as the shock wave compressed the air, heating it to very high temperatures and causing combustible material to simply burst into flames.

Most of the molten magma fountain would have broken up into fragments ranging in size from less than a thousandth of an inch to a yard or more in diameter and would have partially solidified at an altitude of two or three miles. The larger fragments—along with car-sized chunks of the mountain itself—would have fallen back to earth within a radius of three to seven miles. The microfragments, however, would have been carried skyward by powerful convection currents.

As the second phase of the eruption continued, a vast mushroom cloud of ash and debris would have penetrated far into the stratosphere, reaching altitudes of up to thirty miles and carried aloft by extremely strong, high-temperature convection currents, moving at hurricane-force speeds.

In the center of the volcano, temperatures would have reached 1,650 degrees Fahrenheit, generating the heat that forced the ash cloud heavenward. As the mushroom cloud increasingly blotted out the light of the sun and day was turned into night, ash would have rained down on forests and fields alike up to a thousand miles away, and houses would have been shaken by the eruption at similar distances. The sea for dozens of miles around would have been covered with a six-foot-thick floating carpet of pumice, and ships at sea would have become terminally stranded in this volcanic quagmire.

Stupendous amounts of magma, vaporized seawater, and ultrafine hydrovolcanic ash (generated by magma-seawater interaction) would by now have been hurled into the sky, and a substantial percentage of it would have entered the upper part of the earth's atmosphere, the stratosphere. As it spread sideways at high altitude, away from the immediate area of the eruption, the material cooled and the water-vapor component would have then condensed directly into vast clouds of tiny ice crystals. It is estimated that the entire eruption may have generated up to 25 cubic miles of ice crystals; spread out in a thin layer in the stratosphere, these would have caused sunlight diffraction and cooling over vast areas of the globe. Superfine hydrovolcanic ash and huge quantities of sulfur and carbon dioxide gas would have had similar effects. Unlike ordinary volcanic ash, which falls to earth within a few months, hydrovolcanic ash, high-altitude ice-crystal clouds, and sulfuric acid and carbon dioxide aerosols (minute drops) can stay in the stratosphere for years, forming a long-term barrier to normal sunlight and solar heat transmission.

Within hours after the start of the second phase of the eruption, part of the huge mushroom cloud above the volcano would have become too heavy with ash to stay aloft. This part would have collapsed back to the ground, spreading horizontally over land and sea in all directions away from the volcano in what is called a pyroclastic flow, but thousands of times larger than similar flows that partly destroyed the island of Montserrat in the Caribbean in 1997–98.

This horizontally moving cloud would have swept across the ground (and the sea) like a boiling-hot tidal wave of steam, sulfur, air, carbon dioxide, carbon monoxide, ash, and rocks. This hot, poisonous wall of destruction, more than a thousand feet high, would have moved outward perhaps as much as forty miles from the volcano at up to 250 miles per hour, killing anything in its path.

The Third Phase

Then, as the eruption progressed further, the third phase would have begun. Because the huge magma chamber beneath the surface was now partially empty, its roof would have been unable to support the weight of the rock above it. As a result, it would have fallen inward, causing a sudden catastrophic drop of between three hundred and a thousand feet in the level of the land above. As the land surface sank below the level of the adjacent sea, the sea itself would have surged in to cover the former land. Seawater would have again come into direct contact with some of the remaining molten magma, and there would have been a series of immense explosions, producing even larger pyroclastic flows.

The Final Phase

After the catastrophic pyroclastic flow, eruption, and caldera collapse, the fourth and final phase of the eruption would have begun. The explosions would have started to subside over a period of weeks or even months, during which quiet episodes might have persisted for several days or more, punctuated by eruptive bursts of dwindling power. The caldera probably left small island vents that continued to periodically belch steam and ash several miles into the sky for years to come as the residual magma deep below the caldera gradually was quenched.

Comparing this scientific account with the description in the Javanese *Book of Ancient Kings*, we can see that the whole event appears to have been recorded with some accuracy: "At last, the mountain burst into pieces with a tremendous roar and sank into the deepest of the earth. The water of the sea rose and inundated the land. The land became sea and the island [of Java/Sumatra] divided into two parts."

In the past, virtually all geologists thought that the fall in land level could have been caused only by gradual tectonic forces. But a reanalysis of the available geological evidence carried out by volcanologists shows that that view is incorrect. The crucial land-level reduction that caused the formation of the Straits of Sunda *could* have occurred as a result of a volcanic caldera eruption. This geological evidence, when combined with the Chinese historical, Javanese quasi-historical, ice-core, and other evidence, makes the Sunda Straits caldera, proto-Krakatoa, the most likely site of the 535 supereruption.

The Eruption's Significance

The 535 eruption was, as near as can be determined, one of the largest volcanic events of the past fifty thousand years. Whether looked at in terms of short- and medium-term climatic effects, caldera size (assuming proto-Krakatoa was the culprit), or ice-core evidence, the eruption was of truly mammoth proportions. Climatologically, the tree-ring evidence shows that it was the worst worldwide event in the tree-ring record. Looking at the ice cores, we see that it may well have been the largest event to show up in both northern and southern ice caps for the past two thousand years. . . .

The immediate effects of the 535 eruption and a possible second eruption from a different (and as yet unlocated) volcano in

c. 540 lasted five to seven years in the Northern Hemisphere and even longer in the Southern Hemisphere. However, poorly understood climatic feedback systems were almost certainly responsible for years of further climatic instability (including subsequent droughts) in the Northern Hemisphere (up till c. 560) and in the Southern Hemisphere (up till the 580s). The eruption(s), directly and/or through feedback, altered the world climate for decades, and in some regions for up to half a century.

The explosion and climatic changes destabilized human geopolitics and culture, either directly or through the medium of ecological disruption and disease. And because the event, through its climatic consequences, impacted on the whole world, it had the effect of resynchronizing world history.

For the people who lived then, it was a catastrophe of unparalleled proportions. Procopius [a Roman historian], referring to the darkened sun, later wrote that "from the time this thing happened, men were not free from war, nor pestilence nor anything leading to death." However, for us today, the sixth-century catastrophe and the swirling tide of interacting events that flowed from it shed new light on the origins of our modern world, on the processes of history, and—perhaps most alarmingly—on the ultimate fragility of our planet's human culture and geopolitical structure.

Krakatau: 1883

By Robert Decker and Barbara Decker

Krakatau (also known by the name Krakatoa) has been called the most violent eruption ever recorded and may have created the loudest sound in history. Although it is commonly referred to as a single island or volcano, Krakatau is actually a small group of islands located in the Sunda Straits of Indonesia.

This account by Robert Decker and Barbara Decker includes a sea captain's log describing the volcano's first stirrings and its continuation through that evening's nightmarish events. The greatest blast happened the following day when a "fearful explosion" blackened the air and sent a column of rock and fire an estimated fifty miles into the atmosphere. An entire island exploded, and the gargantuan tidal wave, or tsunami, that washed over nearby islands was responsible for most of the thirty-six thousand fatalities. The stupendous eruption of 1883 is said to have begun the era of modern volcanology as all aspects of the eruption have been studied by scores of scientists since then.

Barbara Decker is a professional writer who has written several books and articles about volcanoes with her husband. Robert Decker, past president of the International Association of Volcanology and Chemistry of the Earth's Interior, is currently head of the Geological Survey's Hawaiian Volcano Observatory.

Most of our present mountains, valleys, and plains have been slowly sculptured by erosion over the last million years, and these in turn have formed from other landscapes long vanished. Volcanoes, though, operate on a different time scale.

Krakatau, alias Krakatoa, disgorged 18 cubic kilometers of rock, produced tidal waves over 30 meters high. and formed a circular depression 6 kilometers in diameter and 1 kilometer deep (called a *caldera*) in less than one day. In that short time, Krakatau erupted more volcanic rock than is formed along all the oceanic rifts in one year. The Earth does evolve slowly, but much of the change is the sum of many catastrophic moments.

The explosion of Krakatau in 1883 produced worldwide effects. The noise was heard for thousands of kilometers, and the shock wave recorded on barographs around the world. The dust that was lifted into the stratosphere circled the globe producing astonishing visual effects: colorful sunrises and sunsets and a blue-green appearance of the sun and moon. The dust spread westward, encircling the equator in two weeks, then drifted both north and south. Average solar radiation in Europe decreased 10 percent over the next 3 years, and average world temperatures were below normal.

As attention from many fields focused on the eruption and its aftermath, scientists came to realize the worldwide interdependence of land, sea, and air.

Krakatau is a small group of uninhabited volcanic islands in the Sunda Straits between Java and Sumatra, on a main oceanic trade route between Europe and the Orient. There had not been an eruption there since 1680. The highest peak on the islands was less than 1000 meters tall; it was not an imposing cone.

Eruptions Begin

On May 20, 1883, Krakatau began to erupt with small explosions of ash from a low crater on the north end of the main island. Dutch scientists visited the island on May 27 and noted that much of the vegetation had been killed but not burned by the accumulation of fine volcanic ash. Some blocks of pumice, a rock composed of glassy froth so filled with bubble holes that it floats on water, were also noticed.

The activity declined until late June, when passing ships reported two columns of steam. Apparently a new vent had also become active. On August 11 the island was again visited. Three vents were producing small explosive eruptions, but not much additional ash had accumulated since May, and some trees were still alive on the southern peak of the island.

Things changed on August 26. At 1 P.M. noises like thunder were heard up to 200 kilometers away, and by 2 P.M. a black cloud had climbed to 27 kilometers over Krakatau. The closest witnesses were aboard the British ship *Charles Bal*, bound for Hong Kong. The captain's log gives some vivid impressions of the climax of the eruption.

The Captain's Log

(August 26) At noon, wind west-south-west, weather fine, the Island of Krakatoa to the northeast of us, but only a small portion of the northeast point, close to the water, showing; rest of the island covered with a dense black cloud. At 2:30 P.M. noticed some agitation about the Point of Krakatoa; clouds or something being propelled with amazing velocity to the northeast. To us it looked like blinding rain, and had the appearance of a furious squall of ashen hue. At once shortened sail to topsails and foresail. At 5 P.M. the roaring noise continued and increased, darkness spread over the sky, and a hail of pumice stone fell on us, many pieces being of considerable size and quite warm. Had to cover up the skylights to save the glass, while feet and head had to be protected with boots and southwesters. About 6 P.M. the fall of larger stones ceased, but there continued a steady fall of a smaller kind, most blinding to the eyes, and covering the decks with 3 to 4 inches very speedily, while an intense blackness covered the sky and land and sea. Sailed on our course until we got what we thought was a sight of Fourth Point Light; then brought the ship to the wind, southwest, as we could not see any distance, and we knew not what might be in the Straits, the night being a fearful one. The blinding fall of sand and stones, the intense blackness above and around us, broken only by the incessant glare of varied kinds of lightning and the continued explosive roar of Krakatoa, made our situation a truly awful one. At 11 P.M., having stood off from the Java shore, wind strong from the southwest, the island, eleven miles west-north-west, became more visible, chains of fire appearing to ascend and descend between the sky and it, while on the southwest end there seemed to be a continued roll of balls of white fire; the wind, though strong, was hot and choking, sulphurous, with a smell like burning cinders. From midnight to 4 A.M. the same impenetrable darkness continuing, the roaring of Krakatoa less continuous but more explosive in sound, the sky one second intense blackness and the next a blaze of fire; mastheads and yardarms studded with electrical glows and a peculiar pinky flame coming from clouds which seemed to touch the mastheads and yardarms. At 6 A.M., being able to make out the Java shore, set sail. Passed Anjer at 8:30 A.M., close enough in to make out the houses, but could see no movement of any kind. At 11:15 there was a fearful explosion in the direction of Krakatoa, now over 30 miles distant. We saw a wave rush right on to Button Island, apparently sweeping right over the south part and rising half way up the north and east sides. This we saw repeated twice, but the helmsman says he saw it once before we looked. The same waves seemed also to run right on to the Java shore. The sky rapidly covered in, by 11:30 A.M. we were inclosed in a darkness that might almost be felt. At the same time commenced a downpour of mud, sand, and I know not what; ship going northeast by north, seven knots under three lower topsails; put out the side lights, placed two men on the lookout forward, while mate and second mate looked out on either quarter, and one man employed washing the mud off binnacle glass. We had seen two vessels to the north and northwest of us before the sky closed in, adding much to the anxiety of our position. At noon the darkness was so intense that we had to grope about the decks, and although speaking to each other on the poop, yet could not see each other. This horrible state and downpour of mud continued until 1:30 P.M., the roarings of the volcano and lightnings being something fearful. By 2 P.M. we could see the yards aloft, and the fall of mud ceased. Up to midnight the sky hung dark and heavy, a little sand falling at times, the roaring of the volcano very distinct although we were fully sixty-five or seventy miles northeast from it. Such darkness and time of it few would conceive, and many, I dare say, would disbelieve. The ship, from truck to waterline, is as if cemented; spars, sails, blocks, and ropes in a terrible mess; but, thank God, nobody hurt or ship damaged. On the other hand, how fares it with Anjer, Merak, and the other little villages on the Java coast?

(*Nature*, vol. 29, 1883, pp. 140–41.)

Not well. Over 36,000 people had been drowned by the great
tidal waves generated by the major explosion and collapse. The
tidal waves were the real catastrophe of Krakatau. Ships in the
area were not swamped, but slowly rode up and down with the
swells whose crests and troughs were kilometers apart. But as
these waves reached shallow water, particularly in the bays that
faced Krakatau, they began to curl and break, forming huge surf
waves that swept inland and swamped everything in their reach.
In some locations where they were channeled into narrowing
bays they rose to over 30 meters above sea level.

Effects of the Explosion

The 10 A.M. explosion was the largest natural concussion ever
recorded. The ash cloud rose to 80 kilometers above Krakatau,
and the detonation was heard in Australia. The main tidal wave
swept the coasts of Java and Sumatra about half an hour after the
explosion.

At 10:50 A.M. a second huge explosion occurred but apparently this one did not generate tidal waves. Throughout the afternoon and night of August 27 explosions of diminishing intensity recurred and finally ceased. A few small eruptions
continued into September and October, but most of the violence
was packed into the time between 1 P.M. August 26 and 11 A.M.
August 27.

The Sunda Straits were choked with floating pumice. As ships
began to make their way through, sailors soon noticed that most
of Krakatau was missing. Where the peak of Danan had reached
450 meters high, there was ocean 200 meters deep. A 5- by 8-
kilometer chunk of the main island had disappeared, and a curv-
ing cliff, 800 meters high, cut through the south peak down to
sea level. The origin of this caldera has been a major subject of
debate ever since the 1883 eruption.

Forming Calderas

Some thought the island had simply blown its top and the debris
had scattered into blocks and ashes. However, over 90 percent of
the erupted material was pumice of a texture and composition
not found in the wreck of the old volcano. It must have come
from the magma chamber beneath the volcano; when the volcano had emptied this chamber, the top collapsed into the void
below. The great tidal waves were probably generated by the col-

lapse that followed the largest explosion.

Most geologists now believe that calderas are largely formed by collapse. These great circular basins, up to 20 kilometers in diameter, are common volcanic features and attest to even greater eruptions than Krakatau in prehistoric times.

The Krakatau pumice covered thousands of square kilometers and is calculated to have a volume of 18 cubic kilometers. The volume of island that had disappeared was about 6 cubic kilometers; the difference can be explained by the fact that expanded pumice has three times the volume of its solid rock equivalent. . . .

The Powers of Nature

Australians 4800 kilometers away heard the largest Krakatau explosion as the sound of cannon firing. The shock wave rattled doors and broke windows in west Java, and was recorded on barographs around the Earth. The more sensitive instruments recorded 2 or 3 world-circling trips of the air wave, each circuit taking about 36 hours. The 58-megaton nuclear test in the atmosphere over northern Russia in 1961 produced similar world-circling air-pressure waves.

It is popular to compare the powers of nature with man-made explosions, but this is often misleading. The thermal energy in the 18 cubic kilometers of hot material erupted by Krakatau is equal to that of nearly 5000 megatons of hydrogen bombs. However, only about 5 percent of the Krakatau thermal energy was converted to mechanical energy and its release was spread out over a day's time, in contrast to the almost instant release of energy in a nuclear bomb. Even so, it is somewhat reassuring that nature can still unleash more raw energy than that in man's biggest bomb.

After Krakatau

Quiet prevailed at Krakatau for 44 years; then small explosive submarine eruptions began on the north rim of the caldera in 1927. Soon a small cinder cone was built above sea level; it was named Anak Krakatau—child of Krakatau. This small island has had a fitful history over the past 50 years, sometimes growing during periods of small ash and cinder explosions, and sometimes being nearly washed away by waves from the Indian Ocean. . . .

Will Krakatau ever erupt as violently again? Probably not for many thousands of years. The present composition of the lava indicates that most of the highly explosive magma was expelled in 1883, and new magma of this type forms slowly. However, other volcanoes on the Ring of Fire may be approaching the stage of Krakatau in 1883. Calderas are a common volcanic feature and new ones can be expected to form every few centuries.

Mount St. Helens: Mountain with a Death Wish

BY ROWE FINDLEY

The eruption of Mount St. Helens has been called the worst volcanic disaster in the history of the United States and has generated voluminous research in the field of volcanology. In this article, the first in a series of three classic articles written by Rowe Findley about Mount St. Helens, the author vividly describes the days prior to the eruption as well as the final catastrophic explosion itself. Findley relates his personal experience of being on the lip of the crater just days before the mountain blew. His firsthand report, which may be one of the most detailed accounts ever written about a volcanic eruption, includes Findley's meetings with people living and stationed around the base of the sleeping giant as well as a description of the eruption and its aftermath.

Rowe Findley is a freelance journalist and former assistant editor for National Geographic *magazine.*

First I must tell you that I count it no small wonder to be alive. Looking back on the fateful events preceding Mount St. Helens' terrible eruption last May 18 [1980] I recognize that I—and others—had been drawn into a strange kind of Russian roulette with that volcano in the Cascades.

For many weeks the mountain had masked its potential for tragedy with minor eruptions, then seemed to doze. In our efforts to get a close-range account of a significant geologic event, we moved in with the innocence of the uninitiated—until sudden holocaust shadowed us with peril and changed our lives forever.

The very beauty of the mountain helped deceive us. It was a

mountain in praise of mountains, towering over lesser peaks, its near-perfect cone glistening white in all seasons. Thousands through the years had given it their hearts—climbers, artists, photographers, lovers of beauty's ultimate expression. Some were among the 61 people drawn into its deadly embrace on that shining Sunday morning in May.

For all its splendor, Mount St. Helens was a time bomb, ticking away toward a trigger labeled "self-destruct." Seven weeks before, the world received notice of the mountain's brooding when it first vented plumes of steam and ash. Its immediate domain in southwestern Washington, a favored land of deep forests, rushing streams, rich farmlands, and flourishing cities, waited anxiously as successive eruptions and earthquakes dirtied its crown and fractured its sides.

Then anxieties eased as days and weeks passed without disaster. Though the volcano seethed and trembled, and its bruised north flank bulged morbidly, there were even some who voiced impatience for bigger eruptions. To many, the mountain appeared to be calibrating down toward unreadable calm.

When the Eruption Started

"Vancouver! Vancouver! This is it. . . ." With those words—tinged with excitement rather than panic, hearers said—David Johnston, geologist for the United States Geological Survey, announced the end of calm and the start of cataclysm. Thirty-year-old blond-bearded David was stationed at a USGS camp called Coldwater II, six miles from the mountaintop, to monitor eruptions.

Those words were his last. The eruption he reported was powerful and unexpectedly lateral. Much of the initial blast was nozzled horizontally, fanning out northwest and northeast, its hurricane wave of scalding gases and fire-hot debris traveling at 200 miles an hour. Its force catapulted the geologist and the house trailer that sheltered him off a high ridge and into space above Coldwater Creek. His body has yet to be found.

The start of the eruption has been fixed at 8:32 A.M. Inevitably, the atomic bomb is cited for comparison of magnitude, and the energy computed is that of 500 Hiroshimas. In a quadrant extending roughly west to north, but including a shallower fan to the northeast, 150-foot Douglas firs were uprooted or broken like brittle straws for distances as far as 17 miles from the mountain. An earthquake registering 5.0 on the Richter scale triggered the collapse of the fractured north side of the volcano, which was perhaps a factor in the devastating horizontal venting that followed. Tobogganing on a cushion of hot gases, the disintegrating north wall and cascades of rock swept down over the North Fork of the Toutle River, burying it under as much as 200 feet of new fill, which spread downstream in a 15-mile-long debris flow. The

The powerful 1980 eruption of Mount St. Helens caused billions of dollars in damage and killed dozens of people.

lateral blast hurled a thick blanket of ash over collapsing trees, tumbled bulldozers and logging trucks, crumpled pickups and station wagons, adding to the hopelessness of rescue efforts.

Soon the nozzling of the eruption turned entirely upward, and a roiling pillar of ash thrust some 12 miles into the Sunday morning sky, flanked by nervous jabs of orange lightning. The pillar plumed eastward into a widening dark cloud that would give Yakima, 85 miles distant, midnight blackness at 9:30 A.M. and would last the day. Much of eastern Washington, northern Idaho, and western Montana would be brought to a halt by the ashfall. Within days the silt from the mountain would reach the Pacific, after causing destructive floods on the Toutle and Cowlitz Rivers and closing the busy Columbia to deep-draft ships. By Wednesday the cloud would reach the Atlantic.

I refer to no notes in setting down these events, because they have cut a deep track in my mind. In fact, my memory unbidden replays sequences unendingly, perhaps because of their awesome magnitude and perhaps because they involve a deep sense of personal loss. I have only to close my eyes and ears to the present, and I see the faces and hear familiar names. . . .

Photographer Reid Blackburn

Reid Blackburn. I knew him only a week—the week before the May 18 eruption. At 27 he was a master of cameras and a student of words, a journalism graduate of Linfield College in Oregon and five-year photographer with the Vancouver *Columbian,* a radio technician, a backcountry trekker. He had just the right talents to keep vigil on the volcano and to fire two remote, radio-controlled cameras recording simultaneous images of significant events. For this meaningful project he was on loan from the *Columbian* to the USGS and the National Geographic Society. His post was a mountainside logging-road camp called Coldwater I, eight miles from the crest of Mount St. Helens, three miles farther west than Coldwater II.

Colleagues say that Reid had the incisive eye of the born portrait photographer, capturing a face precisely when the mask falls away to reveal an instant of truth. He was as gifted in filming animals, anticipating the wistful look of a puppy, the trust of a lamb.

Nine months before, Reid had married Fay Mall, a member of the *Columbian'*s office staff, who shared his life's goals and ambitions.

I first met Reid on Sunday, May 11, when I helicoptered to Coldwater I and spent the night there to watch the mountain. I returned the following Wednesday, Thursday, and Friday. The talk ranged from newspapering to backpacking. As we talked on Thursday afternoon, I felt the ground sway like a boat on water. "An earthquake," Reid said without expression. "It's about 4.5." Repeated jolts had calibrated him.

The eight miles that separated Reid from the crater seemed a reasonable margin of safety before May 18. Afterward, with four feet of ash blanketing the camp, and in the knowledge that people twice that far from the mountain had died, I found it hard to think reasonably about margins of safety.

Harry Won't Leave

Harry Truman. He was a man who rejected margins of safety. For more than half a century he had lived at the foot of Mount St. Helens on the shores of Spirit Lake. When sheriff's deputies ordered all residents to leave for safety, Harry said no. Harry had raised the adjectival use of profanity to a new high, and in a position statement that demonstrated his art, he told me why he wouldn't leave:

> I'm going to stay right here because, I'll tell you why, my home and my _____ life's here. My wife and I, we both vowed years and years ago that we'd never leave Spirit Lake. We loved it. It's part of me, and I'm part of that _____ mountain. And if it took my place, and I got out of here, I wouldn't live a week anyway; I wouldn't live a day, not a _____ day. By God, my wife went down that _____ road _____ feet first, and that's the way I'm gonna go or I'm not gonna go.

Harry and his wife, Edna, had built a lodge and cabins by the lake, and their resort became a favored retreat for two generations of vacationers. Three years ago Edna died, and Harry closed the lodge, renting only a few cabins and boats each summer. When a steel gate was placed across the highway, barring outsiders but locking Harry in, he still did not change his mind: "I said block the _____ road, and don't let anyone through till Christmas ten years ago. I'm havin' a hell of a time livin' my life alone. I'm king of all I survey, I got _____ plenty whiskey, I got food enough for 15 years, and I'm settin' high on the _____ hog."

Harry said that he had provisioned an old mine shaft with ample drink and victuals, and many of his friends hope he might yet dig out of such a retreat. But the lack of warning preceding the May 18 eruption makes it all but certain that Harry was caught in or near his beloved lodge, which now lies crushed under thick debris and the raised level of Spirit Lake.

The mountain he elected never to leave rewarded him with an eternal embrace, a cataclysmic burial of a magnitude befitting deity more than man, an extravaganza befitting even Harry's gift for vocal brimstone.

Geologist David Johnston

David Johnston. You already know of his fate, but now you must know of his promise. Like Reid Blackburn's credentials for photography, David Johnston's training for geology was impeccable. The University of Illinois awarded him a degree in geology with honors, the University of Washington conferred master's and doctorate, and the National Science Foundation granted him a fellowship.

Better than most observers, David knew the awesome potential of Mount St. Helens. "This mountain is a powder keg, and the fuse is lit," he said, "but we don't know how long the fuse is." Yet he responded to the need for samples from the crater by volunteering to be the sampler.

"He was a marathon runner in excellent condition," explained Lon Stickney, USGS contract helicopter pilot, who had made more landings on the mountain than any other human. "David figured he could get down into the crater and back out again faster than any of his colleagues."

The Start of a Geologic Event

St. Helens became part of my life late last March. I had been working on a prospective *National Geographic* article on the national forests, and the Fuji-like eminence of Mount St. Helens—named for an 18th-century British diplomat—dominated Gifford Pinchot National Forest. On March 21 my friend Gerry Gause of the Forest Service's regional office in Portland phoned me in Washington, D.C., and said that earthquakes were shaking the mountain. I checked flight schedules and began to read up on Mount St. Helens.

To my hand came an aptly titled USGS paper, "Potential Hazards from Future Eruptions of Mount St. Helens," by Drs. Dwight Crandell and Donal Mullineaux. They said that St. Helens had been the most active of the Cascade volcanoes, and for a quarter century beginning in 1831 had concocted various combinations of steam, ash, mudflows, and lava eruptions. Before the 20th century ended, they predicted, another eruption was likely.

The quakes grew in number and force; the dormant volcano was stretching and stirring. By Wednesday, March 26, I was convinced, and scheduled an early flight Friday. The mountain yawned on Thursday afternoon, venting steam and ash. By the time I arrived Friday morning, intermittent plumes rose two miles above the peak and tinged its northeast slopes sooty gray.

This was the start of a geologic event—the first volcanic eruption in the contiguous 48 states since California's Lassen, another Cascade peak, shut down in 1917 after a three-year run. St. Helens became a siren to geologists, journalists, and the just plain curious who crowded into Portland, Oregon, and into Vancouver, Kelso, and Longview, Washington. Seers competed in foreseeing holocaust, T-shirt vendors had visions of hot sales, and sign makers exhausted plays on the word "ash." Sample: "St. Helens— keep your ash off my lawn." There were some people who irreverently christened the mountain Old Shake and Bake.

The name seemed deserved as late March became April and April slid past with the mountain still not fully awake. A second crater appeared beside the first, then the two merged into a single bowl 1,700 feet across and 850 feet deep. But the eruption level, geologists said, remained "low-energy mode."

St. Helens' Legends

Despite such restraint, there was growing suspense for the country roundabout. It was rugged country, still largely remote except by air, its high places inaccessible under snows most of the year. This was a country of lava caves some thought were home to Sasquatch, or Bigfoot, the giant apelike beast of legend and controversy. This was the wild country over which D.B. Cooper parachuted from a hijacked jet in 1971 with $200,000 in cash; he was never found, though a few thousand of his currency was.

Once this was a land of Indian legend, too, including one in which the favors of a beautiful maiden caused a battle between two rival warriors. They hurled fiery rocks at each other and so angered the Great Spirit that he turned the three into Mount St.

Helens, Mount Hood, and Mount Adams.

Spirit Lake, the mirror for the beauty of Mount St. Helens, owes its name to Indian stories of the disappearance of canoeists on its waters as strange moanings arose.

What would arise from modern-day St. Helens was of immediate concern last spring. The situation put great pressure on geologists for forecasts, but they lacked experience with volcanoes such as St. Helens, a composite of alternating layers of ash and lava. They worked long hours to place instruments on and around the mountain: seismometers to record quakes, gravity meters to gauge vertical swellings, tiltmeters and laser targets to detect outward bulging. A Dartmouth College team led by Dr. Richard Stoiber flew circles around the peak to sample its hot breath. Increased sulfur dioxide content would signal magma on the move. At the University of Washington, at Portland State, at other area universities, faculty geologists monitored their seismographs and analyzed ash samples from the mountain for any clues to its intentions.

Stepping onto the Crater

By Saturday, May 10, the pulse was heavier—some quakes approached 5.0 on the Richter scale. Infrared aerial photos showed several hot spots in the crater and on the flanks. Most alarming of all, the mountain's north face was swelling; it had already bulged laterally by some 300 feet and was still distending at a rate of five feet a day. The volcano would not remain on "hold" much longer.

Still, the third-of-a-mile-wide crater looked drowsy enough in the bright sunlight of late Sunday morning, May 11. With Dr. Marvin Beeson, geochemist at Portland State University, photographer David Cupp and I hopped out of a helicopter onto the crater's northeast lip.

Marvin sought ash samples for analysis. Most of the ash was old, ground-up mountain, but new, glassy ash could be collected and, if it proved high in silica, would indicate how explosive the eruption might be.

Our pilot, Kent Wooldridge, Army trained and Vietnam conditioned, made two precautionary passes before coming to a six-inch-high hover. My jump to the volcano's crest reminded me of watching Neil Armstrong's first step on the moon; would I sink into the mixed ash and snow to my knees or to my hips? Grate-

Ash from the Mount St. Helens eruption affected areas thousands of miles away from the volcano.

fully I found that its consistency was like coarse sand; I sank barely to my ankles, and walking was easy.

While Marvin gathered ash from the crater's lip and David documented the scene on film, I looked around at this uncertain new world. Hundreds of feet below, wispy steam breathed gently from the crater's throat. The south side, towering some 500 feet above us and capped by a disintegrating glacier, constantly whispered and rattled with cascading ice and rock.

The dirty snow was pocked with softball-size holes. With a start I realized that each hole held a rock or ice chunk lately hurled out of the crater. I wondered how good I would be at the volcano's version of dodge ball, I wondered when the next earth tremor was due, and I wondered why Marvin was so slow at spooning samples.

Harry Hanging in There

The week between May 11 and 18 now seems to me part of another life. There was the overnight of the 11th at Coldwater I with Reid. There was time that evening to drive down to Harry Truman's lodge.

Harry greeted me cheerily, iced bourbon in hand, a couple of his 16 house cats scampering underfoot. Yes, his birds had come back—the camp robbers and wrens and blackbirds he fed. Most of them had vanished after the March 27 eruption. The raccoons had never left. The three feet of snow that blanketed his grounds had now melted; long winter was over. He and the mountain were still on speaking terms, and it hadn't told him anything to change his mind about staying.

It was a time for looking back across his 84 years, to his boyhood in West Virginia, to his teenage years in Washington State, where his father had moved to work in the timber, to the Los Angeles of the 1920s, where, Harry said, he used a service station as a front to sell bootleg whiskey that he had brought in by boat from Canada. To years when the late Justice William O. Douglas visited his Spirit Lake lodge, and to his World War II meeting with the other Harry Truman when the latter was the U.S. Vice President.

"By God, if we had Harry S. Truman in Washington now, he'd straighten out those _____ in a hurry!"

Harry R. Truman of Spirit Lake (he never told me what the "R." stood for) had found his life troubled since the mountain began to awaken.

> I'm gettin' letters, hundreds of _____ letters from all over the _____ country. Some of 'em want to save me—somebody sent me a 'Bible for the hardheaded.' I get marriage proposals—now why would some 18-year-old chick want to marry an old _____ like me. I get dozens of letters from children who worry about me.

Schoolchildren and Harry

The children's letters moved Harry, especially a batch from an entire class at Clear Lake Elementary School, near Salem, Oregon. Harry said he planned ultimately to answer all his mail, but he wished he could visit the kids at Clear Lake. "I'd like to explain to them about me and the mountain."

Harry's wish met with enthusiasm at the school, and so a helicopter was arranged to take him there on Wednesday, May 14. I went along, on what proved to be Harry's last trip away from his beloved Spirit Lake.

No Santa Claus ever had a warmer greeting; the entire student body—104 strong—cheered and unfurled crayoned banners (Harry—We Love You) as the whirlybird eased down on the schoolyard turf. Principal Kate Mathews and teacher Scott

Torgeson, whose class had written the letters to Harry, did wel-
coming honors. Harry, foregoing his usual adjectives, admirably
explained how it is to have lived a long, full life, and to have
found a piece of the world as dear as life itself. For each child
who wrote him, he had a signed postcard showing Spirit Lake
and the lodge.

But what would he do if he saw the lava coming for him? "I'd
run," Harry said. The earthquakes worried him more than eruptions, he added, and he had endured a few thousand tremors since
the volcano had started to stir. How did he keep from being
tossed out of bed at night? "I wear spurs to bed," Harry said.

More cheers and waves. The helicopter eases up and out across
sun-dappled fields. The jumping-bean cluster of young well-
wishers shrinks and swings out of sight. A panorama of lush
meadows and woodlands, prosperous towns, and ample rivers
slides beneath Harry's attentive gaze. "What a beautiful country
we got, boys—what a beautiful _____ country," Harry said.

Good-bye, Harry, and good luck.

Last Days Before the Eruption

Thursday, May 15. That famous Northwest weather trick—now
you see it but mostly you don't—plagues efforts to learn what
the volcano is doing. A brief glimpse early in the day shows
hardly a steam plume; then the clouds drop a curtain. We sit by
the chopper at Coldwater I through overtures of alternating
cloud and sun, raindrops and rainbows. The curtain over the
mountain never lifts.

Friday, May 16. The mountain is playing games with us. An
early morning radio message from Coldwater I reports St. Helens
in full view. By the time we get aloft, the curtain is closing. By
the time we reach the mountain, the mountain can no longer be
seen.

Saturday, May 17. All sunshine and no clouds. The mountain
drowses on. The north-face bulge continues—swelling five feet
a day; other signs say that nothing is about to happen. No need
to keep flying around the sleepy mountain.

Instead, I drive to Cougar, a little timber-industry settlement
some 12 miles southwest of the mountain, to see my friends
Mort and Sandy Mortensen, who run the Wildwood Inn, a café
and bar catering to loggers, fishermen, and whoever else turns
up. Lately, business had been hit-and-miss, depending on what

the volcano was doing. The town had been evacuated more than once, and there was no business.

I took the time to reassure myself that the Wildwood Inn's new deep fryer was still turning out delectable fried chicken. And that was the last day of my final week from another life.

The Blast

Sunday, May 18. First sun finds the mountain still drowsing. Because it is drowsing, I decide not to watch it today, a decision that soon will seem like the quintessence of wisdom. Because it is drowsing, others—campers, hikers, photographers, a few timber cutters—will be drawn in, or at least feel no need to hurry out. Their regrets will soon be compressed into a few terrible seconds before oblivion.

Ten megatons of TNT. More than 5,000 times the amount dropped in the great raid on Dresden, Germany, in 1945. Made up mostly of carbon dioxide and water vapor, innocuous except when under the terrible pressure and heat of a volcano's insides and then suddenly released.

That 5.0 quake does it. The entire mountainside falls as the gases explode out with a roar heard 200 miles away. The incredible blast rolls north, northwest, and northeast at aircraft speeds. In one continuous thunderous sweep, it scythes down giants of the forest, clear-cutting 200 square miles in all. Within three miles of the summit, the trees simply vanish—transported through the air for unknown distances.

Then comes the ash—fiery, hot, blanketing, suffocating—and a hail of boulders and ice. The multichrome, three-dimensional world of trees, hills, and sky becomes a monotone of powdery gray ash, heating downed logs and automobile tires till they smolder and blaze, blotting out horizons and perceptions of depth. Roiling in the wake, the abrasive, searing dust in mere minutes clouds over the same 200 square miles and beyond, falling on the earth by inches and then by feet.

The failed north wall of the mountain has become a massive sled of earth, crashing irresistibly downslope until it banks up against the steep far wall of the North Toutle Valley. This is the moment of burial for Harry Truman and his lodge, as well as for some twenty summer homes at a site called the Village, a mile down the valley.

The eruption's main force now nozzles upward, and the light-eating pillar of ash quickly carries to 30,000 feet, to 40,000, to 50,000, to 60,000. . . . The top curls over and anvils out and flares and streams broadly eastward on the winds.

The shining Sunday morning turns forebodingly gray and to a blackness in which a hand cannot be seen in front of an eye.

In the eerie gray and black, relieved only by jabs of lightning, filled with thunder and abrading winds, a thousand desperate acts of search and salvation are under way.

Psychological shock waves of unbelief quickly roll across the Pacific Northwest. In Vancouver and Portland, in Kelso and Longview, and in a hundred other cities and towns, the towering dark cloud is ominously visible.

A phone call from my friend Ralph Perry of the Vancouver *Columbian* sends me outside to gaze at the spectacle. As soon as I can, I get airborne for a better look, and recoil from accepting what I see.

The whole top of the mountain is gone.

Lofty, near-symmetrical Mount St. Helens is no more. Where it had towered, there now squats an ugly, flat-topped, truncated abomination. From its center rises a broad unremitting explosion of ash, turning blue-gray in the overspreading shadow of its ever widening cloud. In the far deepening gloom, orange lightning flashes like the flicking of serpents' tongues. From the foot of the awesome mountain there spreads a ground-veiling pall.

Somewhere down there lies Coldwater I, above the rushing waters of Coldwater Creek and the valley I had left in verdant beauty only 40 hours before.

Averting
Disaster

Monitoring Volcanoes and Predicting Eruptions

BY PETER TYSON

Since early in the 1990s, only about 150 active volcanoes have been monitored, while scientists think at least a thousand need to be watched. In this article, Peter Tyson focuses on techniques that have been developed for monitoring volcanoes and predicting eruptions. Since most volcanoes are preceded by tremors in the earth, seismographs placed at volcanic sites may help predict volcanic activity. Three-dimensional mapping may also help volcanologists to understand what magma is doing under the ground; however, because this technique can be unsafe for scientists, a satellite-based system may be activated soon to automatically and continuously transmit information to observatories. Those and other technologies are needed in order to help scientists to more accurately predict volcanic eruptions and save the lives of those living nearby.

Peter Tyson, the managing editor of Earthwatch *magazine and contributing writer for* Technology Review, *has climbed many active and dormant volcanoes.*

In July 1987 I witnessed my first volcanic eruption. Or, should I say, eruptions. I was on a four-day hike to Semeru, the highest mountain on the Indonesia island of Java. When I first saw the volcano, I was still about 20 miles away, on a ridge. Semeru looked imposing: an ash-colored cone rising alone out of a verdant landscape. Suddenly, within minutes, I saw it erupt. From my distant spot there was no sound and the eruption lasted only a minute or so, but a billowing plume of ash rose high into the azure sky, like a stain entering a body of water. I stood in awe. Not long after I first noticed the plume, which a northerly breeze had pushed off to one side, a second began to ascend. For the next several hours, as the sun went down over Java, I watched the cone spew ash dozens of times with great regularity—about every 17 minutes.

If only all volcanoes erupted so predictably. That they don't has provided a daunting challenge to scientists ever since the Roman author Pliny the Younger, unwittingly assuming the mantle as history's first volcanologist, described in 79 A.D. the devastating eruption of Vesuvius that killed his uncle, Pliny the Elder, and buried Pompeii and Herculaneum.

Lately, population pressure has made answering that challenge more pressing than ever. The 1980s saw more deaths from volcanoes (28,500) than in any other decade since the turn of the century, when three Caribbean eruptions in 1902 alone killed 35,500 people. The reason is that people are living where they shouldn't, on the flanks of volcanoes where the soil is rich and the land for the taking. In 1985, an eruption of Colombia's Nevado del Ruiz killed 23,000 people, burying most in a mudflow that interred the town of Armero. When Ruiz awakened with a much larger eruption a century and a half before, in 1845, only about 1,000 people died. Stanley Williams, a volcanologist at Arizona State University, says he's "confident in a horrible way" that an eruption in his lifetime will kill a million if not several million people. Indeed, the U.S. Geological Survey (USGS) maintains that by the year 2000, half a billion people—1 out of every 12 persons on earth—will be at risk from volcanic eruptions.

The Need for Better Prediction

The United States is not exempt. After Indonesia and Japan, this country contains more active volcanoes than any other. Many of them—55 out of 65—dot the sparsely populated Aleutian chain in Alaska. But 20 of them can be found in national parks, and enough rear over populous regions to keep volcanologists biting their nails. For example, Mt. Rainier, a glaciated volcano that has had eruptions in the past and subsequent mudflows disturbingly similar to those at Colombia's Ruiz, hangs over the Seattle-Tacoma region in northwest Washington. Like the residents of ill-fated Armero, tens of thousands of Seattle residents have built their homes on dried-up mudflows from past eruptions. Even people who feel safely distant from volcanoes may be in danger, for eruptive ash clouds pose a serious hazard to aircraft. In 1982, a jumbo jet en route from Singapore to Australia flew through an ash plume over Indonesia's Galunggung volcano. All four engines lost power and the plane glided for 13 terrifying minutes before the pilots succeeded in restarting the engines. In 1989, a similar occurrence over Redoubt volcano in Alaska resulted in $80 million in damage to the aircraft. When Alaska's Mt. Spurr erupted in 1992, its ash cloud appeared after several days over Ohio, one of the busiest air spaces in the world. Diversions around the visible cloud resulted in costly rerouting delays. "It's only a matter of time before a passenger craft goes down in one of these eruptions," says Charles Connor, a volcanologist at the Southwest Research Institute in San Antonio, Tex.

Yet even as the need for successful prediction has risen, so have the threats to what volcanologists themselves deem an immature and inexact science. The 1985 Ruiz eruption and the devastating Mexico City earthquake the same year prompted the United Nations to declare the 1990s the International Decade for Natural Disaster Reduction. The Decade Volcanoes Project, one piece of the IDNDR's pie, called for more work on predicting eruptions and otherwise reducing risk from the world's most threatening volcanoes.

The project gave renewed hope to volcanologists that their science would finally get a much-needed boost. Yet, says Robert Tilling, a volcanologist with the USGS Volcano Hazards Program, "there has been a lot of talk but little action." Indeed, the picture looks bleak even for the USGS itself, which budget cutters in Congress would like to shut down. Acknowledging his own bias as a USGS employee, Tilling says that the loss of the survey would be "a calamity," because the USGS is the only government agency mandated by Congress to study volcanoes and warn the public of the hazards of eruptions, and it has arguably become the world's leader in the nascent art of predicting eruptions.

Devilishly Complex

While admiring the bubbling lava lake of Hawaii's Kilauea volcano during a visit in the early 1870s, Mark Twain wrote, "I suppose that any one of nature's most celebrated wonders will always look rather insignificant to a visitor at first sight, but on a better acquaintance will swell and stretch out and spread abroad, until it finally grows clear beyond his grasp—becomes too stupendous for his comprehension." Many volcanologists would nod their heads in agreement when speaking of volcanoes and their offspring. For while volcanic eruptions have been documented ever since a prehistoric artist sketched a lava-spewing mountain on a cave wall 6,000 years ago in what is now Turkey, predicting exactly when and how they will blow is still devilishly complex. The enormous variety of volcanoes complicates prediction. (Volcanologists differentiate "prediction," which means giving warning hours to days before an eruption, from "forecasting," which refers to foreshadowing events months to years down the road.) Volcanoes occur at rifts in the ocean floor, above subduction zones where oceanic plates plunge beneath their continental counterparts, and at so-called "hotspots" beneath tectonic plates. Plate shifting causes earthquakes and creates fissures through which magma—molten rock from the earth's interior, which is less dense than solid rock—rises toward the earth's surface, like a cork in water. As the magma nears the surface, the pressure eases and dissolved gases begin to boil out of the molten rock, an expansion that further drives eruptions.

Named after Vulcano, an island off Italy that erupted so frequently that Romans believed it to be the forge of Vulcan, the god of fire, volcanoes have come to mean both any vent in the earth's crust through which magma issues and any resulting landform. These range from mere cracks to soaring peaks like Kilimanjaro and Fuji. As if this diversity weren't enough to keep track of, several of the most destructive eruptions in recent years occurred at mountains not previously thought, even by volcanologists, to be volcanoes. Mt. Pinatubo in the Philippines was such a phantom volcano, and Mexico's El Chichon was considered "just a funny pointed little hill," says Williams, before it exploded in 1982, killing 2,000 people in the worst volcanic disaster in that nation's history.

Volcanologists recognize an equally diverse array of eruption types. While the mechanisms of eruptive activity below ground are believed to be largely the same, once above ground, eruptions tend to do their own thing. "Each volcano tends to have a personality unto itself, each has its own particular behavior," says Tilling. "So it's difficult to extrapolate from detailed studies at one volcano to other volcanoes." The spectrum ranges from Icelandic eruptions that feature lava lazily spilling out of fissures, to explosive eruptions—known as Plinian after the Roman historian—that send so much ash into the atmosphere they can alter global climate. Explosive eruptions are eclipsed only by so-called caldera-forming eruptions. These truly earth-shaking blasts explode with such Herculean force that they leave behind vast, basin-like depressions—calderas—that can stretch 10 or more kilometers across. The largest caldera-forming eruptions, which fortunately have not occurred in recorded-history, make the explosive eruption of Washington's Mt. St. Helens in 1980 seem like a firecracker. In contrast to the cubic kilometer of magmatic phlegm St. Helens coughed up, for instance, large caldera-forming eruptions like the one that carved out Yellowstone some 2 million years ago can expel 500 or more cubic kilometers of material.

To complicate matters, volcanologists identify new species of eruptions all the time, even among closely watched volcanoes. Williams of Arizona State says volcanologists were caught off guard by the violent lateral explosion that blew out the side of Mt. St. Helens in May 1980, the most devastating U.S. eruption. "Once that example existed in our minds, we go around the world and what do we see? Hundreds of examples like St. Helens. They're everywhere. We just hadn't been thinking of them."

Further, some volcanoes can change their behavior with each eruption. For centuries Japanese volcanologists have studied eruptions at Unzen, a volcano on the island of Kyushu, yet they failed to foresee the terrible pyroclastic flows that killed 43 people in June 1991, including three volcanologists. Pyroclastic flows—also known as *nuee ardente*, French for "glowing clouds"—are heavier-than-air mixtures of superheated volcanic ash and gas that, propelled by the rapid expansion of gases bubbling out of magma, race downhill at speeds often exceeding 160 kilometers per hour, incinerating everything in their path. One of the scientists killed during the 1991 eruption of Unzen was Maurice Krafft, who was making a public-service film on eruptive dangers with his wife Katia when the flow overwhelmed them. Shortly before his death, Krafft put the prediction problem succinctly when he wrote that "we still have no general theory of volcanism that might allow us at last to know precisely why a volcano erupts." As if unknowingly passing the baton, he added, "A formidable task awaits future generations of volcanologists."

Can We Predict Eruptions?

In a word, yes. But that assertion, like saying we can predict the
weather, bears significant conditions. Volcanologists can predict
eruptions—if they have a thorough understanding of a volcano's
eruptive history, if they can install the proper instrumentation on
a volcano well in advance of an eruption, and if they can continuously monitor and adequately interpret data coming from that
equipment. But even then, like their counterparts in meteorology,
volcanologists can only offer probabilities that an event will occur; they can never be sure how severe a predicted eruption will
be or, for that matter, whether it will even break the surface.

Still, under ideal conditions, volcanologists have recently met
with a great deal of success in foretelling eruptions. While they
were caught off guard by the exact timing and magnitude of the
1980 Mt. St. Helens eruption, for example, their timely warnings
of an impending blow prompted the U.S. Forest Service to evacuate people from dangerous areas near the volcano. Though 57

THE ERUPTION
ALERT SYSTEM

During the eruption of Pinatubo in 1991, scientists
and military personnel developed a system to rate
the activity of the volcano. This rating system, similar to
weather alert systems, was created to warn the public of
the dangers of the volcano.

Level 1: Low level of unrest; no eruption imminent.
Level 2: Evidence of magma; could eventually lead to
          an eruption.
Level 3: Increasing unrest; eruption possible in two
          weeks.
Level 4: Intense unrest with a long period of earth-
          quakes; eruption possible in twenty-four hours.
Level 5: Eruption under way.

Allison Lassieur, *Volcanoes*. San Diego: Lucent, 2002.

people died in the eruption, perhaps 20,000 lives were saved, says
William Rose, a volcanologist at Michigan Technological University. Similarly, a USGS team rushed to the Philippines' Mt.
Pinatubo in the spring of 1991 and successfully augured the June
eruption, leading to evacuations that saved thousands if not tens
of thousands of lives and millions of dollars worth of military
equipment at the nearby Clark Air Force Base.

Predicting Kilauea's Eruptions

Not surprisingly, volcanologists have had the most success at volcanoes that host their own observatories. In 1912, Thomas A.
Jaggar, head of the Geology Department at MIT, founded the
first volcano observatory in the United States on Kilauea. (There
are now two others—one based in Anchorage, Alaska, and another in Vancouver, Washington, near Mt. St. Helens.) Over the
succeeding decades, researchers at the Hawaiian Volcano Observatory developed many of the techniques used today and can
now predict Kilauea's eruptions to a tee.

They know when and how Kilauea will erupt because it does
so frequently and predictably, and because after decades of in-
tensive study they know the volcano inside and out. Learning as
much as possible about a volcano's previous behavior is the essential first step in anticipating future blows, just as knowing a career criminal's record can help indicate what he might do next.
"There is no doubt that the eruptive history of the volcano is the
main key for long-term prediction," says Yuri Doubik, a Russian
volcanologist who has studied past eruptions on the Kamchatka
Peninsula for 35 years. Such work entails laboriously picking
through the physical remains of previous eruptions. And mapping such old lava flows, pyroclastic deposits, and other volcanic
debris distributed around a crater can reveal much about the timing, type, direction, and magnitude of previous blows.

Satellite data can greatly aid such mapping and volcanologists
are looking forward to using images generated by the Earth Ob-
serving System. . . . The satellite's purpose is to study environmental
ills such as global warming and depletion of the ozone layer, but
it will also gather information of use to volcanologists, such as on
gas concentrations in the atmosphere over volcanoes and images
clear enough to reveal the fallout from former eruptions.

When a volcano's eruptive history is known, researchers can
more confidently turn to modern techniques to help them call

the next eruption. The most valuable among these, volcanologists agree, is monitoring a volcano's seismicity—the frequency and distribution of underlying earthquakes. Use of the seismologist's tool in volcanology has come a long way since Frank Perret, one-time assistant to Thomas Edison, gleaned the frequency of the small shocks that continually shake Vesuvius's flanks by biting down on the metal frame of his bed, which was set in cement. Today sophisticated seismographs can register the magnitude, escalation, and epicenters of earthquakes that occur as magma moves beneath volcanoes. The more seismographs technicians deploy on a volcano, the more complete picture they get of the mountain's plumbing.

Seismic networks can transmit data by radio 24 hours a day to computer-equipped monitoring stations well out of harm's reach, enabling scientists to safely watch for changes in "nature's noise," as one volcanologist labeled the geophysical status quo within a volcano. Computer-based seismic data acquisition and analysis systems, which in essence constitute portable observatories, enabled the USGS's Volcano Crisis Assistance Team to successfully augur the 1991 eruption of Mt. Pinatubo by revealing the build-up of long-period quakes, a type of volcanic tremor that often appears just before and during eruptions.

Three-Dimensional Mapping

While seismicity is the workhorse, monitoring ground deformation is another up-and-coming technique that allows three-dimensional mapping of what's occurring underground. Magma rising from the depths often pushes the skin of a volcano up and out, like a balloon filling with air. Sensitive tiltmeters [instrument for measuring angles of slopes] and surveying instruments can measure and record the slightest changes, which help volcanologists determine, for example, roughly how deep a magma source is, how fast it is moving, and where on a volcano it might erupt. Such monitoring has helped scientists anticipate eruptions at Hawaii's Kilauea and Mauna Loa volcanoes, which deform in predictable ways and at predictable rates.

One drawback is that ground deformation requires scientists to climb volcanoes to take measurements—a perilous undertaking. But USGS volcanologists are now testing a prototype of a fully automated ground-deformation system that takes advantage of the U.S. Navy's satellite-based Global Positioning System,

Scientists install a tiltmeter, designed to measure and record changes in the slope or "tilt" of the ground.

which can automatically and continuously transmit information on latitude, longitude, and elevation, with a resolution of a few centimeters or less, to remote observatories. . . .

Japanese Boxes

Monitoring of volcanic gases got its start in the 1950s when enterprising Japanese researchers put beakers of potassium hydroxide, a strong, basic solution, on Honshu's Asama volcano, which was beginning to show signs of erupting. As the highly acidic gases released by the crater seeped through holes in a crate covering the beakers, they increasingly altered the solution's composition in the months before a large eruption.

Today, volcanologists use so-called "Japanese boxes" routinely, though again they must check the beakers manually. To surmount this problem, Williams, who was nearly killed during a small but deadly eruption while visiting the crater of Colombia's Galeras volcano in January 1993, is designing an electronic Japanese box that will automatically and continuously transmit data to a remote observatory. About the size of a briefcase, the battery-powered unit has tiny electrochemical sensors that create currents proportional to the amounts of various volcanic gases in the air. . . .

While volcanologists feel confident that these ever-improving technologies will enable them to predict when an eruption is

about to occur, they still cannot reliably estimate an impending
eruption's size or exact nature. How large will the eruption be?
Will it be explosive like St. Helens or effusive like Kilauea? Indeed, will it even open a vent in the surface? To be able to answer
such questions, Tilling and his USGS colleague Peter Lipman ar-
gued in a 1993 article in *Nature* for the need to develop "rugged,
reliable real-time systems" to measure changes not only in seismicity, ground deformation, and gases, but also in gravitational and
electromagnetic fields—in short, equipment to read the gamut of
signals given out by a restless volcano. "There's no magic bullet in
predicting volcanic eruptions," Connor says. "The key thing is to
cross-correlate as many different observations as possible."

Tilling says volcanologists also need to get a better handle on
the basic mechanisms behind precursory signals, such as the long-
period earthquakes that often precede eruptions. This holds especially true for large caldera-forming eruptions, which have not
occurred since the dawn of civilization. In the mid-1980s, three
volcanic fields believed to hold the potential for one of these
monumental eruptions—California's Long Valley, Papua New
Guinea's Rabaul, and Italy's Campi Flegrei—"turned on" almost
simultaneously, throwing the volcanological community into a
bit of a frenzy. But all three centers calmed down without further ado. Tilling, for one, is confident that this type of eruption
will not come unheralded. "No volcano is going to suddenly
produce one of these humongous eruptions without giving a lot
of signals," he says. "But what will those signals be?"

An Understudied Risk

Beyond improving prediction technologies, the foremost task,
volcanologists agree, is to monitor more volcanoes. Maurice
Krafft noted in the early 1990s that the world's roughly 30 volcano observatories keep tabs on only 150 active volcanoes, while
"we should be monitoring a thousand." Roughly 1,500 volcanoes have erupted in the past 10,000 years and therefore should
be considered active, according to the Smithsonian Institution.
Many dangerous volcanoes lie unwatched, most notably in developing countries such as Ecuador and Guatemala that have no
volcanic monitoring programs under way.

Getting a rudimentary seismic network up and running would
require only a few tens of thousands of dollars per volcano, says
Tilling. Compared with other global research projects, he main-

tains, the effort could be mounted at relatively modest cost. Possible sources of funding are the United Nations, other international funding agencies such as the World Bank, or even the airline industry, which has a vested interest in knowing when volcanoes might explode. "An engine on a 747 probably costs $50 million," says John Dvorak, a USGS volcanologist. "The airlines are a multibillion-dollar industry, and a million here and a million there [for volcano monitoring] isn't going to break them."

Installing instruments on the world's highest-risk volcanoes, most of which lie in the developing world, is not enough, however. Peter Mouginis-Mark, a volcanologist at the University of Hawaii, says that too often developing countries lack scientists or technicians with the expertise to properly interpret data coming from sophisticated equipment loaned by foreign colleagues. Moreover, he adds, looting of the computer equipment is so pervasive that in some cases local authorities have had to hire armed guards.

Need More Funding

Developing countries therefore need help from the developed world, Rose says, not only in technology transfer but also in training programs for local volcanologists. Grant guidelines often limit such outreach, however. The National Science Foundation and NASA, the principal funding bodies in this country for non-USGS volcanologists, have long pushed their applicants to focus on basic rather than applied research. Many volcanologists, particularly those who have witnessed the worst that volcanoes can dole out, including the loss of colleagues, would like to focus more on practical benefits. "It's just that nobody pays us," says Rose, to explain the benefits of the technology to the world at large or how it should be applied. In the same way, funding arrangements often don't encourage volcanologists to do international work, such as helping Ecuador develop a volcano hazard program.

Rose complains that "we're never allowed to spend any money until there's a crisis." Volcanology has been called the Cinderella science that only marches forward on the ashes of catastrophe. After the 1980 Mt. St. Helens eruption, for example, funding for prediction research at the USGS increased tenfold, leading to near-perfect predictions through the mid-1980s of so-called dome-building eruptions within the blown-out crater. But time heals all wounds, and that funding has steadily waned. Says Rose: "I know that there are people who, from the point of view of the strategy of forecasting, would dearly love to have another domestic eruption, because of the opportunity to develop these techniques that only seems to happen when property or people suffer. It's one of the cruel ironies of volcanoes."

The Public Needs Warning

But all the fancy techniques for predicting eruptions are as naught unless the public obtains sufficient warnings. The 1985 eruption of Nevado del Ruiz precipitated such a horrific tragedy because local authorities did not heed the repeated warnings of volcanologists, who only a month before the disaster had published a hazard map foretelling with uncanny accuracy the very mudflows that buried Armero. And while Mexico's Popocatepetl, a volcano 70 kilometers from Mexico City, has recently given volcanologists warnings that it could enter a much more destructive phase at any time, says Williams, Mexican authorities are not taking action.

Educating the public about the dangers from eruptions can make all the difference. When Papua New Guinea's Rabaul volcano showed signs of erupting in 1994, Williams says, "50,000 people got up and walked out of town, even though they weren't told to do so by the official government scientists." Widely distributed hazard maps, chalk boards with up-to-the-minute volcano bulletins, and simulated evacuations encouraged local people to take their own initiative. When the volcano erupted on September 19 of that year, fewer than a dozen people died. "The death toll could have been many thousands, because about 75 percent of the houses collapsed," says Williams, who was on the scene. "It's a wonderful example of how people can be educated to save themselves."

A month after I summited Java's Semeru, I climbed Agung, the highest mountain on neighboring Bali. I camped high on the volcano's shoulder so I could make it to the top in time for the sunrise. I woke at 2 A.M. and, stowing my tent off the trail, worked my way up by the light of a headlamp. As I sat on the crater rim, the first light of dawn pushed darkness away and revealed a scene of sublime beauty. Low-lying clouds hid most of the island below, while to the east I could clearly see the island of Lombok, with its own soaring volcano. Soon the sun broke the horizon, casting the crater rim in a striking amber light. The volcano could not have seemed more at peace with itself. Yet I knew better than to trust appearances. Only a quarter-century before, on the very day in 1963 when Hindu priests led a once-a-century prayer to Agung in Besakih temple on the volcano's southern flank, the mountain had suddenly blown its top, killing 1,148 Balinese and burying thousands of hectares of rich farmland in an ashen sarcophagus. When it comes to sleeping giants like Agung, can we afford to rest easy?

Educating and Alerting the Public

By Sandra Blakeslee

Sandra Blakeslee, an award-winning science writer for the New York Times, *reports that volcanoes are becoming more dangerous because a greater number of people are living in closer proximity to them. Even when scientists warn the public of imminent danger, she writes, their warnings may be ignored. As a result of volcanic disasters, such as the one that killed twenty-six thousand people in Colombia in 1985, scientists are trying to increase the public's awareness of the dangers of volcanoes by means of public education, research, and improved prediction technology. They chose to concentrate their study on fifteen volcanoes in densely populated areas, holding workshops with scientists and local relief workers. In addition, they produced a video that graphically depicts the destruction that volcanoes can cause. In 1991, immediately before the eruption of Mount Pinatubo in the Philippines, this video was rushed to the area and shown on television. The next day fifty thousand people evacuated the area. The volcano erupted several days later, and it is believed that tens of thousands of lives were saved by the evacuation.*

P lanet Earth is planning some spectacular volcanic fireworks and millions of people worldwide are going to have ringside seats.

The next show could start anytime, almost anywhere. It might be Mexico City, where a 17,000-foot volcano named Popocatepetl is spewing ash and poisonous gases toward 20 million homes; it conceivably could explode with the force of 10,000 atomic bombs. Or it might be Vesuvius, the famed volcano that looms over Naples and surrounding Italian towns, home to 11 million people. A small volcano called Soufriere Hills is erupting now [August 1997] on the lush Caribbean island of Montserrat, and has already driven most of the island's people from their homes.

There are about 1,500 active volcanoes, not counting hundreds more under the oceans, and any of them could erupt at any time, said Dr. Tom Casadevall, western regional director of the United States Geological Survey in Menlo Park, California. Of the 1,500, 583 have exploded within the last 400 years, making them particularly dangerous. Each year, scientists observe 50 to 60 volcanoes in various stages of eruption, some gently extruding lava like red hot toothpaste down hillsides, others heaving molten rock particles and noxious gases many miles up into the atmosphere.

The number of people living on the sides of volcanoes and in the valleys below has skyrocketed, said Dr. Stanley Williams, a volcanologist at Arizona State University in Tempe. At least 500 million people live dangerously close to volcanoes, he said. Many dwell in megacities in Asia and Latin America—Tokyo, Manila, Jakarta, Mexico City, Quito—or in cities of at least a million people. Here in the United States, the people of Seattle and Tacoma live in the shadow of Mount Rainier, a 13,000-foot volcano whose mudflows have swept through the places where the cities are situated.

People have been drawn to volcanoes for centuries because the surrounding soils are rich and old volcanic mudflows make nice flat areas for settlement, Dr. Williams said. As population rises and land gets scarcer, the problem is getting worse.

Most of the time, the people who colonize danger areas do not know any better. And the people who do know better, scientists and civil disaster officials, "are not always listened to," said Dr. Grant Heiken, a volcanologist at the Los Alamos National Laboratory in New Mexico.

For example, scientists issued a warning when a high volcano, capped with ice, began rumbling in the mountains of Colombia in 1985. On November 13, the icecap exploded above the town of Armero. The eruption melted snow fields that picked up debris and went roaring down the side of the volcano toward the villages 30 to 40 miles away. The residents were warned that night that a large volcanic mudflow was on the way, Dr. Heiken said. "But it was raining," he said. "People said, 'Why worry, the volcano is far away.' They had only to walk 100 yards to a hill to be safe. That night, 26,000 people died."

*In 1985 mudflows from a volcanic eruption destroyed the town of Armero,
Colombia, killing twenty-six thousand people.*

The International Decade of Natural Hazard Reduction

Scientists were horrified, said Dr. Chris Newhall, a volcanologist at the United States Geological Survey at the University of Washington in Seattle. This episode and other natural disasters prompted the United Nations to declare the 1990's the International Decade of Natural Hazard Reduction, he said. "The notion was, look, the world population is growing, the hazards are not getting any less," he said. "People are moving into marginal lands that are more prone to disasters—volcanoes, floods, earthquakes and hurricanes. The basic idea was to encourage countries to take a hard look at the hazards that their populations were facing and to undertake projects to try and reduce risks."

But the United Nations did not have money for the program, Dr. Newhall said. The International Decade of Hazard Reduction has existed in name only.

What Scientists Are Doing

So the scientists began taking action on their own. Under the auspices of the International Association of Volcanology and Chemistry of the Earth's Interior, "we volcanologists got together and scratched our heads for ideas," Dr. Newhall said. "We came up with three."

First, they made a video that depicts what volcanoes can do to people and property, with such horrifying accuracy that it is not recommended for children under 15. It is being shown to mayors and other public officials in charge of getting people to evacuate when volcanoes threaten to explode.

Second, the scientists picked 15 volcanoes around the world to study intensely. These so-called decade volcanoes are near large population centers and could erupt any time. Workshops have been held at most of them, bringing together scientists and disaster relief officials from the local regions.

Third, there has been an effort to make better predictions of when volcanoes will erupt, using new scientific instruments and insights.

Although real progress has been made, volcanologists face a couple of intractable problems, Dr. Williams said. One is the tendency for people to deny danger even when it is obvious. Also, once a threat is passed, they tend to dismiss it. "They forget that grandma once told a story about how her grandmother was killed by a volcano," he said. And second is the sheer perversity of volcanoes. They may show all the signs and symptoms of erupting and then quiet down, leading the public to accuse scientists of "crying wolf."

Dr. Williams himself nearly lost his life in the Galeras volcano in Colombia while surveying it for the decade project a few years ago.

About Volcanoes

Volcanoes arise from the forces of plate tectonics, and usually occur where great slabs of the Earth's crust are shoved deep into the interior and melted. This molten rock is buoyant, Dr. Williams said, and makes its way back to the surface where it finds a weak spot and explodes to the surface in a volcano.

Most volcanoes live for tens of thousands of years and can remain dormant for centuries between eruptions. The varying amounts of ash, lava and other particles they emit are measured in cubic kilometers, cubes about five-eights of a mile on each side. The 1980 Mount St. Helen's eruption in Washington State, which devastated the countryside, was a small event by volcanic standards, producing only one cubic kilometer, Dr. Williams said.

By contrast, a volcano in Yellowstone, Wyoming, has over the last two million years spewed out 3,800 cubic kilometers of ash and pumice. If Yellowstone produced another huge eruption, it could shut down the country, he said. Airplanes could not fly, trucks could not deliver food and farmers could not grow crops.

Study of the Fifteen Decade Volcanoes

The 15 decade volcanoes are being studied, Dr. Newhall said, "to arrive at a synopsis of what was and was not understood about how each volcano works. Also, what questions need to be addressed to make future forecasts more accurate in terms of the size and timing of eruptions."

First, geologists look all around each volcano, mapping ash deposits and mud flows to determine the nature of each past eruption, how far it went and, with carbon dating, how long ago it happened, said Dr. Richard Fisher, a professor emeritus of geology at the University of California at Santa Barbara. Then they map out hazard zones. If a volcano produced serious damage in one place, it can do so again, he said.

Next, the scientists study the low-frequency earthquakes that these volcanoes tend to produce. Such earthquakes make a special noise that is related to growing magma domes, the ominous bulges formed by molten rock making its way to the surface. With almost every volcano, people in the area begin feeling quakes, hearing noises and smelling gases for days, weeks or months before an eruption, Dr. Williams said. They often sense it is coming.

Using a variety of instruments, the scientists are measuring the size and growth of magma domes at decade volcanoes and the gases they emit. Steam and carbon dioxide indicate that high pressures are building up. Sulfur dioxide emissions "tell you that the magma is in communication with the atmosphere," Dr. Fisher said. It is a definite danger sign.

Warning Signs

A few things send volcanologists running for the hills, Dr. Newhall said. If a volcano starts to produce low harmonic tremors, a steady hum of seismic waves, for tens of minutes or hours, it is time to flee. It means magma is rising up the conduit and building gas pressure. Recent studies on decade volcanoes also point to another forewarning: sulfur dioxide levels sometimes drop right before the big blast.

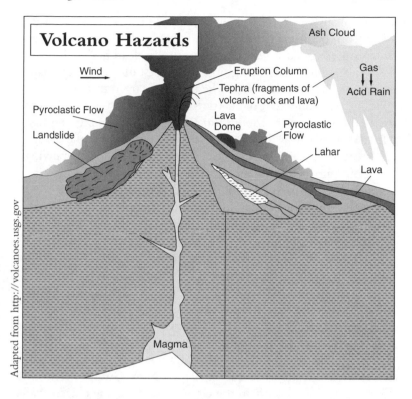

The trouble is, other volcanoes do not give these warning signs before they explode. "You do the best you can," Dr. Newhall said. "You look at spatial and temporal changes and you plot them for weeks, months and then you make your best shot at interpreting what is going on."

The major cause of death in volcanoes is not hot lava or rivers of mud, but rather glowing clouds of superhot gas and ash particles that silently sweep down the volcano's flank and across the countryside at 60 miles an hour, vaporizing everything in their path. These pyroclastic flows can knock down stone walls 10 feet thick and have killed thousands of people in less than two minutes, he said.

In the movies, people outrun the flows, but in real life, the flows desiccate the flesh and fry the lungs of everyone in their path. Ninety-eight percent of the people in Guatemala live on the surface of a pyroclastic flow that raced over the countryside 75,000 years ago, Dr. Williams said.

From watching volcano movies and films of the rather gentle

and atypical volcanoes in Hawaii, people think they can walk
away from danger, Dr. Heiken said. The volcanologists' video
shows otherwise. It is very blunt and shows dead bodies, he said.
"When people see it, they gulp and say, 'Could that really happen here? How far did you say the town was from that volcano?'"

In 1991, a rough cut of the newly made video was rushed to
the Philippines, where Mount Pinatubo was threatening to erupt.
The day after it was shown on television, 50,000 people evacuated voluntarily. A few days later, the volcano erupted, spewing
12 cubic kilometers of material. "We are convinced that the
video saved tens of thousands of lives," Dr. Heiken said.

Getting the word out and convincing people to evacuate is a
huge challenge, Dr. Williams said. Ultimately, whether people live
or die depends as much on communication as on science. "We
try to teach people how not to freak out," he said. "They think
falling ash is lethal, but it's not. The problem is that the ash is
three times heavier than water so their roof can collapse in hours.
We teach people to get under the strongest table in the room,
near the corner. The air is full of static electricity, so radios and
traffic lights go haywire and they get scared."

Such public education saves lives, Dr. Heiken said. In September 1994, a volcano at Rabaul in Papua New Guinea destroyed
75 percent of the homes in the city. But because the citizens had
been trained in evacuation procedures, they did not panic and got
away safely. Only five people died.

It is a different story in the United States "where any form of
government message is taken as a challenge to people's rights to
do whatever they want," Dr. Casadevall said. At Mount St.
Helens, some people refused to evacuate and died as a result.
Around Mount Rainier, 250,000 people live on the surface of
volcanic mudflows that are less than 500 years old. Some older
flows reached Seattle. These people are being told they should be
ready to move to high ground if Mount Rainier shows signs of
erupting, he said. The problem is that with some towns, there is
one narrow road out over a single bridge. "We try to make sure
people have this information," Dr. Casadevall said, "but if they
refuse to leave, they're on their own."

Preventing
Volcanic
Catastrophe

BY JOHN W. EWERT, THOMAS L. MURRAY,
ANDREW B. LOCKHART, AND C. DAN MILLER

---

*In this selection, the authors focus on the U.S. International Volcano
Disaster Assistance Program (VDAP) and its goals and strategies. After
thousands of people were killed by a major volcanic disaster at Nevado
del Ruiz in Colombia in 1985, the U.S. Geological Survey (USGS)
created VDAP to help reduce loss and disruption of life in areas of volcanic activity.*

*VDAP is a team of volcanologists at the Cascades Volcano Observa-
tory (CVO) in Washington State with a supply of volcano-monitoring
equipment that can be quickly dispatched to areas in volcanic danger.
VDAP helps scientists to respond to dangerous situations anywhere in
the world while continuing to improve the monitoring capabilities of observatories.*

*Geologist John W. Ewert works for the USGS at the Cascades Vol-
cano Observatory. Thomas L. Murray works for the Air Force Research
Laboratory's Directed Energy Directorate.*

*Andrew B. Lockhart is a geophysicist who works for the USGS at
the Cascades Volcano Observatory. C. Dan Miller is a volcanologist with
the USGS and chief of the VDAP.*

W hen the seismograph began to record the violent
earth-shaking caused by yet another eruption of the
Nevado del Ruiz volcano in Colombia, no one
thought that a few hours later more than 23,000 people would
be dead, killed by lahars (volcanic debris flows) in towns and villages several tens of kilometers away from the volcano. Before the

John W. Ewert, Thomas L. Murray, Andrew B. Lockhart, and C. Dan Miller,
"Preventing Volcanic Disaster: The U.S. International Volcano Assistance
Program," *Earthquakes & Volcanoes*, vol. 24, no. 6, 1993, pp. 270–91.

fatal eruption the volcano was being monitored by scientists at a seismic station located 9 km from the summit, and information about the volcano's activity was being sent to Colombian emergency-response coordinators who were charged with alerting the public of the danger from the active volcano. Furthermore, areas known to be in the pathways of lahars had already been identified on maps, and communities at risk had been told of their precarious locations.

Unfortunately, a storm on November 13, 1985, obscured the glacier-clad summit of Nevado del Ruiz. On that night an explosive eruption tore through the summit and spewed approximately 20 million cubic meters of hot ash and rocks across the snow-covered glacier. These materials were transported across the snow pack by avalanches of hot volcanic debris (pyroclastic flows) and fast-moving, hot, turbulent clouds of gas and ash (pyroclastic surges). The hot pyroclastic flows and surges caused rapid melting of the snow and ice and created large volumes of water that swept down canyons leading away from the summit. As these floods of water descended the volcano, they picked up loose debris and soil from the canyon floors and walls, growing both in volume and density, to form hot lahars. In the river valleys farther down the volcano's flanks, the lahars were as much as 40 m thick and traveled at velocities as fast as 50 km/h. Two and a half hours after the start of the eruption one of the lahars reached Armero, 74 km from the explosion crater. In a few short minutes most of the town was swept away or buried in a torrent of mud and boulders, and three quarters of the townspeople perished.

After the fatal eruption, volcanologists of the U.S. Geological Survey were dispatched to Colombia to quickly establish a seismic and tiltmeter [instrument for measuring angles of slope] network at the volcano and to help Colombian scientists assess the likelihood of future eruptions and lahars. At the same time, a painstaking search began for the circumstances that led to the disaster.

A Preventable Disaster

It soon became clear that no single factor was responsible for the disaster. Contributing factors were a lack of a timely hazards evaluation (a hazard map took nearly a year to complete after the first signs of volcanic unrest and was available for distribution only days before the eruption), an inadequate monitoring system at the volcano, and ineffective procedures for communicating information and making decisions during the emergency. In hindsight, the disaster at Nevado del Ruiz could have been prevented.

The realization that disasters like that at Nevado del Ruiz might be prevented launched the Volcano Disaster Assistance Program (VDAP) in August 1986. With support from USAID [United States Agency for International Development] through its Office of Foreign Disaster Assistance, the U.S. Geological Survey created VDAP to assist developing countries during volcanic crises. During its short existence, VDAP has assisted Ecuador, Colombia, Guatemala, the Philippines, and other countries to reduce the loss of life and property from volcanic eruptions and to prepare for future volcanic crises. The successful response to the 1991 eruption of Mount Pinatubo in the Philippines stands as VDAP's most extraordinary contribution to volcano-hazard mitigation.

Goal and Strategy of VDAP

The goal of the program is to reduce loss of life and minimize economic disruption in countries that experience volcanic eruptions. Drawing from the lessons of the Nevado del Ruiz disaster, the strategy to meet this goal includes the following elements:

- Develop the capability to rapidly deploy a volcano-monitoring network anywhere in the world.
- Assist local scientists in geologic studies to assess volcano hazards.
- Work with local scientists to interpret monitoring data and to disseminate hazard information.
- Train local scientists to use monitoring techniques to forecast volcanic eruptions.

The ability to respond rapidly with volcano-monitoring equipment is not sufficient by itself to mitigate volcano hazards. Although an early scientific response to a reawakening volcano is critical to making reliable forecasts of the timing and nature of future eruptions, hazard mitigation is most effective when volcanoes have been monitored for many years before eruptive activity begins. To assist local scientists, VDAP provides the monitoring equipment and personnel needed to quickly establish an effective volcano-monitoring network. Data gathered by the network helps volcanologists to forecast eruptive activity and issue timely eruption warnings.

The record of a volcano's past eruptive activity is preserved in the volcanic rocks and unconsolidated volcanic deposits that surround the volcano. Evaluation of a volcano's past eruptions by dating these deposits and determining their mode of origin, provides information about the nature of possible future eruptive activity and associated hazards at the site. The results of these studies, summarized in volcano-hazard assessment and hazard-zonation maps, help public officials to prepare land-use maps and determine risk during emergencies.

The program has the best chance of success when VDAP and host-country scientists have worked together for a period of time before volcanic unrest becomes a crisis. Thus, VDAP scientists are cooperating with scientists from other countries to help them prepare for future volcanic activity in their countries. Workshops and training programs for participants from developing countries are conducted in the United States and in the host countries. To meet the needs of the many Spanish-speaking host countries, USGS publications about volcano hazards and volcano monitoring are published in Spanish as well as English.

What Is VDAP?

The principal components of VDAP are: (1) a core team of five volcanologists at the Cascades Volcano Observatory (CVO) in Vancouver, Washington, and (2) a cache of volcano-monitoring equipment also kept at the observatory.

The fields of expertise of the core team include geology, geophysics, hydrology, and electronics. The team plans daily operations, purchases and develops equipment, and participates in all VDAP responses. Other scientists, from within and outside the USGS, supplement the core group as needs arise. The second component is a complete cache of monitoring equipment that functions as a portable volcano observatory. Much of the equipment was developed or modified by the USGS; with the essential requirements that it must be durable, relatively inexpensive, and easily transported. The monitoring equipment can be set up quickly and is self-contained.

In addition, the VDAP staff continually strives to improve hardware and software systems to enhance the monitoring capabilities of volcano observatories.

In addition to saving lives from volcanic eruptions in other countries, collaborative work by VDAP has significantly strengthened the ability of the USGS Volcano Hazards and Geothermal Studies Program to respond to future volcanic crises within the United States. By working with VDAP, USGS scientists have gained additional experience in monitoring active volcanoes and communicating volcano-hazard information to emergency management officials and people living in hazardous areas. Participation in volcano emergencies abroad has provided critical field tests for newly developed or modified volcano-monitoring instruments. Furthermore, these collaborative experiences have led to the development of data-acquisition and data-analysis systems that run on widely available personal computers (PCs). In the event of a crisis at a volcano in the United States, the USGS has the ability to supplement an existing network or install an entirely new one in a matter of days. This capability did not exist when Mount St. Helens became active in 1980 or during the 1985 crisis at Nevado del Ruiz.

Another consequence of VDAP is the extent to which volcano-monitoring systems are standardized and widely distributed in both developed and developing countries. Use of standardized equipment makes it possible for Central and South American countries to cooperate in installing, maintaining, or exchanging components of the system and in interpreting data in familiar formats.

VDAP's Portable Volcano Observatory

Volcanologists monitor changes in the physical or geochemical state of a volcano induced by magma movement beneath the volcano. Movement of magma generally causes swarms of earthquakes and produces other types of seismic events, swelling or subsidence of a volcano's flanks, and sometimes changes in the amount or types of gases emitted from a volcano. By monitoring these phenomena, volcanologists are sometimes able to forecast eruptions days to weeks ahead of time and to detect remotely the occurrence of certain volcanic events such as explosions or lahars.

Monitoring techniques have vastly improved due to the recent advances in electronics and the development of the personal computer. A portable volcano observatory is now possible because of the large storage capacity and rapid data analysis that can be accomplished with a PC. The use of PCs makes it possible for scientists to quickly establish a complete monitoring network at a restless volcano with a nearby data-gathering and data-analysis base station. This has proven extremely important for geologists conducting field work at restless volcanoes, both for their personal safety and for correlating field observations in real time with geophysical data so as to effectively update hazard assessments during rapidly changing conditions.

Volcanic Explosivity Index

The Volcanic Explosivity Index (VEI) is a scale volcanologists use to measure the magnitude of volcanic eruptions.

VEI	Description	Plume Height	Volume	Classification	How Often	Example
0	non-explosive	<100 m	1000s m^3	Hawaiian	daily	Kilauea
1	gentle	100–1000 m	10,000s m^3	Haw/Strombolian	daily	Stromboli
2	explosive	1–5 km	1,000,000s m^3	Strom/Vulcanian	weekly	Galeras, 1992
3	severe	3–15 km	10,000,000s m^3	Vulcanian	yearly	Ruiz, 1985
4	cataclysmic	10–25 km	100,000,000s m^3	Vulc/Plinian	10's of years	Galunggung, 1982
5	paroxysmal	>25 km	1 km^3	Plinian	100's of years	St. Helens, 1981
6	colossal	>25 km	10s km^3	Plin/Ultra-Plinian	100's of years	Krakatau, 1883
7	super-colossal	>25 km	100s km^3	Ultra-Plinian	1,000's of years	Tambora, 1815
8	mega-colossal	>25 km	1,000s km^3	Ultra-Plinian	10,000's of years	Yellowstone, 2 Ma

APPENDIX

The Most Disastrous Eruptions

This list presents estimates of the numbers of people killed by some of history's deadliest volcanoes. Some volcanoes were larger than these but resulted in fewer than five hundred fatalities. In other cases, there is insufficient data with which to estimate the numbers of people killed.

Deaths	Volcano	Year	Major Cause of Death
92,000	Tambora, Indonesia	1815	Starvation
36,417	Krakatau, Indonesia	1883	Tsunami
29,025	Mt. Pelee, Martinique	1902	Ash flows
25,000	Ruiz, Colombia	1985	Mudflows
14,300	Unzen, Japan	1792	Volcano collapse, tsunami
9,350	Laki, Iceland	1783	Starvation
5,110	Kelut, Indonesia	1919	Mudflows
4,011	Galunggung, Indonesia	1882	Mudflows
3,500	Vesuvius, Italy	1631	Mudflows, lava flows
3,360	Vesuvius, Italy	79	Ash flows and falls
2,957	Papandayan, Indonesia	1772	Ash flows
2,942	Lamington, Papua N.G.	1951	Ash flows
2,000	El Chichon, Mexico	1982	Ash flows
1,680	Soufriere, St. Vincent	1902	Ash flows
1,475	Oshima, Japan	1741	Tsunami
1,377	Asama, Japan	1783	Ash flows, mudflows
1,335	Taal, Philippines	1911	Ash flows
1,200	Mayon, Philippines	1814	Mudflows
1,184	Agung, Indonesia	1963	Ash flows
1,000	Cotopaxi, Ecuador	1877	Mudflows
800	Pinatubo, Philippines	1991	Roof collapses and disease
700	Komagatake, Japan	1640	Tsunami
700	Ruiz, Colombia	1845	Mudflows
500	Hibok-Hibok, Philippines	1951	Ash flows

GLOSSARY

aa: A type of lava flow that has a rough, fragmental surface.

active volcano: A volcano that is erupting or has erupted in recorded history.

ash: Fine particles formed in explosive eruptions.

bomb: A still-viscous lava lump ejected from an explosive eruption that develops a rounded shape while in flight.

caldera: A gigantic depression in the earth formed when a volcano collapses into its underground magma chamber.

cinder: A pyroclastic fragment that is about one centimeter in diameter.

cinder cone: A steep conical hill formed by the accumulation of cinders and other loose material expelled from a volcanic vent by escaping gases.

composite volcano: A steep-sided volcano composed of layers of volcanic rocks, which usually include high-viscosity lava and fragmented debris such as lahar and pyroclastic deposits.

conduit: The pipe or crack through which magma moves.

crater: A bowl- or funnel-shaped depression, generally in the top of a volcanic cone; often, the major vent of volcanic products.

dome: A mound of lava extruded onto the surface of a volcano.

dormant volcano: A volcano that is not presently erupting but is considered likely to do so in the future.

dust: The finer particles of volcanic ash.

eruption cloud: A gaseous cloud of volcanic ash and other pyroclastics that forms from a volcanic explosion.

extinct volcano: A volcano that is not erupting and is not expected to do so in the future; a dead volcano.

fault: A fracture in the earth's crust along which there has been movement.

flank eruption: An eruption from the side of a volcano (in contrast to a summit eruption).

fume: A gaseous cloud without volcanic ash.

hot-spot volcanoes: Volcanoes related to a persistent heat source in the mantle.

lahar: A volcanic mudflow.

lava: Magma or molten rock that has reached the surface; the resulting solid rock after cooling.

lava lake: A lake of molten lava in a volcanic crater or depression.

magma: Molten rock beneath the surface of the earth.

magma chamber: An underground reservoir in the earth's crust filled with magma.

mantle: The zone of the earth below the crust and above the core.

mudflow: A thick mixture of water and debris that can be very fast moving.

nuee ardente: A fast-moving, dense "glowing cloud" of hot volcanic ash and gas ejected from a volcano.

pahoehoe: A type of lava flow with a smooth, billowy, or undulating surface.

plate tectonics: The theory that the earth's crust is composed of large independently moving plates.

plume: A rising column of magma from deep in the earth's mantle; the heat source for hot-spot volcanoes.

pyroclastic flow: A very hot mixture of volcanic ash, pumice, rock fragments, and gas that can move at hurricane speeds.

Ring of Fire: The regions of mountain-building earthquakes and volcanoes that surround the Pacific Ocean.

shield volcano: A gently sloping volcano in the shape of a flattened dome, built by flows of very fluid basaltic lava.

strato-volcano: A steep volcanic cone built by both lava flows and pyroclastic eruptions.

subduction zone: The zone of convergence of two tectonic plates, one of which usually overrides the other.

tephra: A general term for all airfall pyroclastics from a volcano.

tsunami: A great sea wave produced by a submarine earthquake or a marine volcanic eruption.

vent: An opening in a volcano through which volcanic material is ejected.

volcano: A vent in the surface of the earth through which magma erupts; the landform constructed by erupted material.

Units of Measurement

cubic meter (m³): 1.3 cubic yards

cubic kilometer (km³): 0.24 cubic miles

kilometer (km): 0.62 miles

meter (m): 3.28 feet

Books

Isaac Asimov, *How Did We Find Out About Volcanoes?* New York: Walker, 1981.

Jelle de Boer and Donald T. Sanders, *Volcanoes in Human History: The Far-Reaching Effects of Major Eruptions.* Princeton, NJ: Princeton University Press, 2002.

Jon Erickson, *Quakes, Eruptions, and Other Geologic Cataclysms.* New York: Facts On File, 1994.

R.V. Fodor, *Earth Afire! Volcanoes and Their Activity.* New York: W. Morrow, 1981.

Peter Francis, *Volcanoes: A Planetary Perspective.* New York: Oxford University Press, 1994.

S.L. Harris, *Fire Mountains of the West: The Cascade and Mono Lake Volcanoes.* Missoula, MT: Mountain Press, 1988.

Maurice Krafft, *Volcanoes: Fire from the Earth.* New York: Harry N. Abrams, 1993.

G.A. MacDonald, *Volcanoes.* Englewood Cliffs, NJ: Prentice-Hall, 1972.

David Ritchie, *The Ring of Fire.* New York: New American Library, 1981.

Kerry Sieh and Simon LeVay, *The Earth in Turmoil.* New York: W.H. Freeman, 1998.

T. Simkin and Richard S. Fiske, *Krakatua: The Volcanic Eruption and Its Effects.* Washington, DC: Smithsonian Institution Press, 1983.

Dick Thompson, *Volcano Cowboys: The Rocky Evolution of a Dangerous Science.* New York: St. Martin's Press, 2000.

Time-Life Books, *Volcano: Continents in Collision*. Alexandria, VA: Time-Life Books, 1982.

Jane Walker, *Fascinating Facts About Volcanoes*. Brookfield, CT: Millbrook Press, 1995.

Periodicals

Henry S.F. Cooper Jr., "Upwardly Mobile," *Natural History*, April 2001.

Robert C. Cowen, "Preventing Pompeii," *Christian Science Monitor*, January 2001.

Cornelia Dean, "Questions of Access in a Land of Ancient Volcanoes," *New York Times*, November 2000.

David Ewing Duncan and Jon Miller, "Volcanoes," *Life*, June 1996.

Michael Durand and John Grattan, "Effects of Volcanic Air Pollution on Health," *Lancet*, January 2001.

Rene Ebersole, "Sleeping Giant," *Current Science*, January 2002.

John W. Ewert et al., "Preventing Volcanic Disaster: The U.S. International Volcano Assistance Program," *Earthquakes & Volcanoes*, vol. 24, no. 6, 1993.

Peter Francis and David Rothery, "Remote Sensing of Active Volcanoes," *Annual Review of Earth and Planetary Sciences*, 2000.

Tim Harris, "When Glaciers Erupt," *Earthwatch: The Journal of Earthwatch Institute*, August 2000.

Nicola Jones, "How Volcanoes Are Running Rings Around the World," *New Scientist*, November 2000.

David Joshua, "Whole Lava Love," *Fortune*, July 2001.

Rainer Kind and Stephan V. Sobolev, "Cutting Edge," *Times Higher Education Supplement*, October 2000.

Jeffrey Kluger and Dan Cray, "Volcanoes with an Attitude," *Time*, February 1997.

Howard LaFranchi, "Living in the Shadow of 'El Popo,'" *Christian Science Monitor*, November 2000.

Michael D. Lemonick and Patrick E. Cole, "Dante Tours the Inferno," *Time*, August 1994.

Jack McClintock, "America's Most Dangerous Volcano," *Science World*, November 2000.

Sid Perkins, "Scientists Analyze Volcanoes' Killing Ways," *Science News*, January 2001.

Paul Simpson, "Ash Thursday," *Focus*, October 1998.

Gar Smith, "The Environmental Catastrophe That Changed History," *Earth Island Journal*, Summer 2000.

Priit J. Vesilind, "Once and Future Fury," *National Geographic*, October 2001.

George P.L. Walker, "Volcanology Came of Age in the Past Two Decades," *Earthquakes and Volcanoes*, 1990.

Tim Weiner, "Watchful Eyes on a Violent Giant," *New York Times*, January 2001.

Karen Wright, "A Volcano Is Born," *Discover*, December 2001.

Websites

Alaska Volcano Observatory (AVO), www.avo.alaska.edu.
   The AVO is a joint program using federal, state, and university resources. The site offers volcano photos, information on volcanology and remote sensing, and weekly updates on monitoring activity.

Cascades Volcano Observatory (CVO), http://vulcan.wr.usgs.gov.
   The CVO is operated by the United States Geological Survey (USGS). It has volcano news and current events, answers to frequently asked questions, hazard maps, photos, useful links, and instructions on how to order USGS maps, reports, and aerial photos.

Electronic Volcano of Dartmouth College, www.dartmouth.edu.
   This site, available in seven different languages, provides information on active volcanoes with maps, photos, and full dissertation texts.

Hawaiian Volcano Observatory (HVO), http://hvo.wr.usgs.gov.
The HVO site, sponsored by the USGS, has information on
Hawaiian volcanoes and volcanic hazards, maps, press releases,
a photo gallery, and access to *Volcano Watch,* a weekly news-
letter written by HVO scientists.

Michigan Technological University's Volcanoes Page, www.geo.
mtu.edu.
This site is sponsored by the Keweenaw Volcano Observatory and was one of the first volcano websites. It provides
scientific and educational information on volcanoes and tries
to provide information that other sites do not have.

Smithsonian Institution Global Volcanism Program, www.volcano.
si.edu.
This website of the National Museum of Natural History
offers volcanic activity reports, information on volcanoes
around the world, and volcano-related products.

U.S. Geological Survey (USGS) Volcanoes Page, http://volcanoes.
usgs.gov.
This site has updates on volcanoes in the United States and
worldwide, volcano fact sheets, and information on volcanic
hazard programs, reducing volcanic risk, and USGS work
abroad.

Volcanoes.com, www.volcanoes.com.
This website is "student friendly," offering a photo gallery,
volcano stories, books and videos, information on recent volcanic activity, and volcano links.

Volcano World of the University of North Dakota, http://volcano.
und.nodak.edu.
Called the Web's premier source of volcano information, the
site has details on current eruptions, volcano adventures, in-
terviews, and listings and descriptions of volcanoes.